To Sarah

may your
horticultural
endeavours
blossom!

Gardening for Gorillas

Trials, tricks, and triumphs of a zoo horticulturalist

Stephen Butler

PREFACE

After I retired in the summer of 2018, I started writing; no rush, gathering thoughts and memories. I retired while Dublin Zoo was a commercial success and deeply involved in animal conservation and education. A modern, forward-looking zoo – unrecognisable from the one I had joined thirty-seven years earlier. Little did I or anyone else, realise what was ahead of the world. The Corona virus, Covid-19, drastically affected us all, with numerous tragic fatalities.

My thoughts are with anyone affected by Covid-19 in whatever way.

I'm a horticulturalist, so my thoughts especially go out to colleagues, many of whom have been temporarily laid off or made redundant. With little or no visitors, many gardens or tourist sites that need gardeners, have had to cut their losses. Gardeners have continually been an easy target. Surely the plants can look after themselves for a while… Time will tell, but the work of getting planting back under control, where control is desired, will be harder for some time.

For Frances, Mark and Susan, with apologies for all the stops on every walk whenever a plant caught my eye.

CONTENTS

INTRODUCTION

Stephen Butler has been an important figure in the gardening community for decades. His knowledge is encyclopaedic, fuelled by a robust curiosity that drives him to investigate the world of plants and horticulture from many angles. In this book, he discusses with relish the practical, the scientific, the historic and – especially – the idiosyncratic aspects of his work.

In his thirty-seven years at Dublin Zoo as head gardener and curator of horticulture, Stephen helped to transform the landscape radically. His tenure coincided with the hatching of a new masterplan for the zoo, where animal welfare became paramount and there was enough funding to develop new naturalistic environments. During his time, the place metamorphosed from a spartan residence for animals into an exciting and beautiful sanctuary. The sterile Victorian pens and the largely flat and barren landscape have been replaced by spacious habitats and a varied terrain of hills, water, winding paths and lush vegetation.

The planting for the different areas is geographically themed, so that it matches the look and spirit of the animals' various homelands. Visitors feel that they are on a voyage of discovery. On the 'Kaziranga Forest Trial,' for example, you wend your way along a sinuous path carved through a jungle of bamboo, magnolia and other luxuriant greenery. Turn corner after corner and you are eventually rewarded by the sight of Asian elephants sunbathing, foraging and frolicking. Elsewhere, you glimpse gorillas across a swampy body of water in a habitat modelled on a rainforest clearing in the Republic of Congo. In the Irish climate, the planting must be pseudo-equatorial rather than the real thing, but jumbo leaves (including phormiums, catalpa, ligularia, gunnera and arum lily) give the right impression of exuberant abundance. The 'African Savanna,' where rhinos, giraffes, zebras, oryx and ostriches roam, is all heat and dust, with spare-looking, drought-tolerant plants. On the raised perimeter, safe from nibbling animals, Australian acacias and other savanna-style plants grow.

Gardening at a zoo presents unique challenges, as Stephen so eloquently describes in this book. Certain plants have to fill so many roles that they are heroic multitaskers. For instance, those that are situated within the animal areas must be sturdy, resilient species that offer shelter and privacy while also doing no harm through poisons or other hazards. If they can offer browsing opportunities without being obliterated, a trunk to climb or some other kind of interest – all the better. Finding species that can satisfy all these criteria is a painstaking job, one for an expert plantsman with an enquiring mind and a vast knowledge. Stephen Butler possesses these characteristics.

He has another attribute that makes all the difference between a workaday planting scheme and one that sings its heart out. He loves plants. He glories in their individual personalities, their relationships with each other and the animal kingdom, their history, their folklore and – so importantly – their aesthetic qualities. The many plant communities that he has designed throughout the zoo perform their duties admirably for the animals and their habitats, while also wowing humans. The excitement is mighty when you come upon a plantation of giant Himalayan lilies, each one a stout three-metre stem hung with multiple outsize flowers; or the two-metre, red-rocket inflorescence of the biennial Echium wildpretii, endemic to the Canary Islands; or a sprinkling of broad-leaved helleborine, a native Irish orchid.

Stephen's passion for all manner of plants has led to a tremendously varied and interesting landscape. The numerous, healthy ecosystems in the thirty-hectare plot now provide a haven for an abundance of wildlife. The diversity in the zoo is greater than that of the surrounding Phoenix Park – Europe's largest enclosed public park in a capital city.

The adventure of greening the zoo, with all its tears and triumphs, is recounted here and threaded with invaluable lessons. Stephen shares his expertise freely and entertainingly. Soil compaction has never been so interesting, nor the range of mulch-dwelling fungi. He includes a helpful section on botany, where sticklers for accuracy will be thrilled to discover that the blackberry is not a berry, but an aggregate fruit composed of many drupelets. A banana, meanwhile, is a berry. This last gem is just one of many that I will cherish in this treasury of knowledge.

Jane Powers
Garden and Nature Writer
February 2021

THANKS

No large garden can survive without help from outside contractors and suppliers. Tree work needs specialised training and machinery too. This was very ably provided by *Shaw Tree Services* and *Arborist Tree Surgeons*. *Dignam Garden Services* always had great machinery for grass care and other work. *Reel-Tech* kept our machinery running. Many years of excellent topsoil from several different suppliers was the foundation for years of growth. Plants came from a variety of nurseries, wholesale or retail, depending on the project. It was always a pleasure to order within Ireland, supporting the great nursery skills here.

Trees and shrubs came from *Yellow Furze*, *Ravensberg*, *Future Forests*, *Flannerys* and *Annaveigh*. Bamboos came from *Stam's Nursery*. Herbaceous plants from *Kilmurry*, *Mt. Venus*, *Leahys*, *Leamore* and *Costins*. Liners came from *Nightpark*. *Irish Water Plants* always had great waterside plants.

Design by Nature supplied excellent Irish origin wildflower seeds. Some of the above have closed unfortunately. Garden supplies and much advice came from *NAD* and *Whelehans*. There was always a steady trickle of advice, plants and supplies from other sources – too numerous to list here – and all were gratefully received at the time. Inspiration often came after visiting other zoos or gardens and talking to like-minded people, particularly at the National Botanic Gardens, Glasnevin and Kilmacurragh and Trinity College Botanic Garden.

Plant people are unbelievably generous with time and often plants too!

DEDICATION

To gardeners everywhere and everywhen. Obviously to the many staff and students involved in the Dublin Zoo horticulture effort over the years. Especially to the original Dublin Zoo team in my early years there – Tom, Christy, Pat, Billy, Simon and Fred. Each of whom probably reckoned I was a trifle mad to try some of the planting we did...but they did it anyway!

PREAMBLE

I was born in West London, within sound of Twickenham Rugby Ground. I was blessed with a family home immediately behind which was a linear park that followed the River Crane. The park was most famous for an old shot tower where round 'shot' was made. Molten lead was poured from the top of the tower into the water below and it made small spheres on the way down. As a kid, I roamed the entire length of the park, all three miles. An uncle of mine, David Snoddy, a soft-spoken gentleman Scot, was the head parkkeeper.

Housing estates (many pre-World War 2) were built along the boundaries of the park, with their back gardens against the boundary, often with the garden overflowing a tad onto the riverbank. The area of the park nearest our house had been used as a bomb damage dump during WW2. It filled in a swampy area and created two separate branches of the river, joining again after an allotment site where we enjoyed a plot for a few years.

Good numbers of willow *Salix spp* and alder *Alnus glutinosa* lined the riverbanks and of course, sycamore *Acer pseudoplatanus*. Outside the estate, on slightly higher ground, there was a small public golf course, with a different range of plants in rough, poor grass and heathland. The golf course had plenty of birch *Betula pendula*, lots of meadow plants and lots of different butterflies too, blues especially.

There were areas of woodland, rough grassland, woodland-edge areas, river edges that had been timbered to stop erosion, which was not so good for wildlife. But for me, the old bomb-damaged dump area was the best. Flattened out, it had become very rough with grass, trees and odd plants, including garden escapes brought in with the bomb damage rubble. I found bladder senna *Colutea arborescens* there. There must have been twenty acres of perennial nettles – the food plant of peacock, red admiral and small tortoiseshell butterflies, with hundreds of larvae there. My parents had planted an ice plant *Hylotelephium spectabile* (*Sedum* in old money) too near the front garden gate. It made a flat doormat sized magnificent smorgasbord of food for nectar-seeking flying insects. The adult butterflies – maybe fifty at a time, including painted ladies (that's another butterfly for anyone entomologically challenged) – would land on this to feed in season, only to be disturbed by anyone opening the gate. Whoever came in, be it the postman, breadman, rent collector – they would always pause on the way back out to look at the butterflies, before reluctantly disturbing them by opening the gate again. It was wonderful.

Within the park there were a few clumps of bamboo *Pseudosasa palmata*. The leaves made toy boats to sail down the river. There was quite a few old pedunculate oak *Quercus robur* trees. These were a reminder of the history of the area. It had been an estate originally, Fulwell Park. While in exile for about twenty years before he died in 1932, this had been the home of the last King of Portugal, Manoel II. The estate road names were a reminder too; Portugal Gardens, Lisbon Avenue, Manoel Road, Augusta Road (for his wife). Manoel's mother had been born not far away, a little bit down the Thames, in York House. In 1914, he held a garden party at Fulwell Park, attended by Queen Victoria and the Empress Marie of Russia. He was active socially in the area, interested in local affairs and supported local churches and the local Hampton Garden Society. He had been an interesting character, no doubt.

This range of habitats, albeit very heavily altered by the hand of man over the years, gave a rich fauna that often visited us in the garden. We had butterflies, bees and hoverflies by the dozen. Chasing these, we had occasional dragonflies and birds galore. I distinctly remember watching an owl being mobbed by other birds while he sat completely ignoring them in the oak tree in our garden. Bats visited at night too. One day an awful noise had us looking up, only to see a heron being harassed by two crows. One crow had caught one of the heron's trailing feet in its beak, causing the heron to give off very distinctive croaks. If a heron could ever be said to sound disgruntled, that was it. Grey squirrels were an absolute pest, especially when chased into the house by a neighbour's cat – up and down the curtains, etc. and back out again. All you had time to do was sit and try to follow the action. Great fun, though Mum was not so happy...

Despite being reared in a busy city, I was very lucky to be surrounded by various habitats with diverse, if somewhat depleted flora and fauna. I fell into being interested in plants. My parents gardened. The unusual plants I found and looked up to identify certainly got me hooked. Trips with family or school to Hampton Court Palace and Gardens five miles away or to the Royal Botanic Gardens at Kew five miles the other way, whetted my appetite for more. I think there was also an element of plant hunting being easier than chasing small brown unidentifiable birds or worse, moths. Both were labelled as SBJs – small brown jobs – and left to flutter by. The interest in general wildlife stayed with me though. I should probably have become a biologist rather than a horticulturalist, but there were no apprentice biologist positions around. The school career advisory teacher, after looking up horticulture, pointed me that way and I went into the Royal Parks apprenticeship scheme, based in Hampton Court Palace just down the road. It was an incredible start to any gardener's career.

The three-year apprenticeship in Hampton Court Palace with the next two years as a Grade 1 Gardener, was followed by the three-year student course at the Royal Botanic Gardens at Kew. This gave me both a great grounding in good horticulture and a permanent fascination with all things plant; from their evolution and their role in atmosphere formation and climate control, to their place in diverse habitats around the planet, our uses for them, their incredible ability to maintain and restore habitats, plus their inherent beauty and wonder. I can seldom look at a plant without thinking where it comes from, how ancient is it, how did it get here, what insects live on it, are there any links to specific animals or insects, can I have one and how do I propagate it.

Taking up a position as head gardener in a zoo was a completely novel idea to me. I had never considered the horticultural aspects of keeping animals on public display. My memories of visiting zoos were as part of school trips to either London Zoo or Whipsnade Zoo. I fear I paid too little attention to the plants, though my main memory is of fences. Yet again, I was blessed to arrive at Dublin Zoo. Although a city zoo and an old one at that – it opened in 1831 a few miles from Dublin city centre, it is in a singularly wonderful setting. It is surrounded by Phoenix Park, one of the largest walled deer parks in Europe, with an 11km wall enclosing 707 hectares. From within the zoo, the only man-made structure you can see outside is the nearby radio mast in an army barracks. A small stream, a mix of surface drainage and springs, flows through a shallow valley. In Victorian times, it was dammed to give a chain of lakes. For many years, the first was in the President of Ireland's estate, *Áras an Uachtaráin*, now largely part of the expanded zoo.

The zoo itself originally had one lake; the larger part was for geese and a smaller part was kept for ducks. Lower down the valley within Phoenix Park, The People's Gardens uses the same water to create a pleasant garden around another small lake. The climate is generally benign, usually without extremes of cold or heat, yet with Ireland's famous all year-round gentle rain. The average summer highs are no more than 25°C and at worst winter lows of -5°C for a few days. That said, cold winters can occur. 1982, 2009 and 2010 had snow and days of -15°C. Unfortunately, many good plants were lost in those cruel winters. Summer drought can occur, as in 1976 and 2018. Though temperatures seldom get over 30°C, the drought may only last a few weeks. In Ireland, an absolute drought is any fourteen consecutive days with less than 0.2mm of rain per day. It can be very wet at times also, but it is more likely to be a gentle, constant dampness that causes problems for plants, especially in winter, rather than extreme rainfall.

After thirty-seven years of zoo horticulture experience, I am far more aware that, unfortunately, a lot of people suffer from 'plant blindness.' This is the inability to notice plants in their environment, often with the same reaction to the importance of plants generally in the biosphere and for the future of *Homo sapiens* too. When it comes to plants, it seems the genus 'homo' (human) is not necessarily always very 'sapiens' (Latin for wise). Plants are so important to every aspect of our lives. They range from the obvious food, shelter and fuel from timber and medicine for so many of our ailments, to the less obvious in daily life but equally important roles in water management, soil conservation and importantly for fauna, habitat creation. Plants are often the last consideration in many designs, whether in zoos, housing, offices or schools. There may be some planting factored in, but is it done with natural habitats in mind that will also allow fauna to thrive? How many mown grass areas, even in part, could be wildflower meadows with associated insects? How many tree plantings could be underplanted with shade-tolerant shrubs for a more complete habitat, with the classic woodland edge that is always full of interest? Equally, is the management of the area carried out with the correct attitude to wildlife, the ecosystem that has developed and the benefits of keeping it as biologically diverse as possible?

Almost all creatures on this unique planet need plants, either as habitat, for food, for shelter or as the actual structure they live on. No plants? Then, no animals. Including humans.

Plants uniquely use the energy of the sun to take carbon dioxide from the air and water from the soil to grow, giving out oxygen as a by-product. This process is called photosynthesis. There are some unusual plants with no roots that take moisture from the air and plants that have no leaves as they are parasitic on the roots of other plants. They are all a wonderful, interconnected mix. Conserve the natural habitat for any large animal and you keep the entire ecosystem with it – plants, essential fungi, animals, insects, algae (especially in the sea) and not forgetting all those friendly bacteria there too.

In the early ages of experiment 'Planet Earth,' it took a billion years for blue-green algae (cyanobacteria – a very primitive early plant-like organism) to release enough oxygen through photosynthesis before air-breathing lifeforms could evolve and plants have generally kept the earth in balance since – not too much carbon dioxide, nor too little oxygen. Humans are rapidly destroying that balance. As a species, we are too many for the planet to sustain. Plus, far too many people suffer to some degree from the aforementioned plant blindness. In fact, it is difficult to avoid being part of the problem. The same can be said for climate blindness too. Our overriding consumerism takes far more than the planet can give. The result being plants pay the price – if not removed completely, then by drastic reduction in diversity, with a similar reduction in fauna too. Many plants can come back from seed or propagation and start rebuilding habitats, animals are far more likely to need difficult and expensive conservation and reintroduction. There are about 400,000 different species of plants, depending on classification systems. The number of mammals is about 4,000, which are the main component of most zoos. Usually they have plenty of movement, compared to reptiles. It is always a challenge to get people really interested in most insects, yet everything has its place in nature.

Many mammals have the most fascinating links to particular plants for pollination or seed distribution. Often the animals need a diverse plant community to do well themselves. Sometimes the plants need a particular animal without which it cannot be pollinated or maybe the seed will not germinate unless it passes first through the gut of only one animal species. There are examples where plants and animals are so linked that the loss of one will threaten the other and that works both ways. The interplay between plants and animals or insects, is a fantastically complex subject, with the most incredible links having evolved through the aeons. There is always something more to discover and so much more to be wondered at.

I hope my efforts to keep the zoo environs densely planted have given people a little more to think about when viewing animals, whether they be in cages, enclosures, exhibits, habitats, zoos, wildlife parks, bio-parks or conservation parks. There is nearly always a way to get more plants into any given area. Though you may need to think very much outside the box for how to do it and make it succeed. Depending on the area's climate, once you satisfy the basic needs for any plant – a large enough area, suitable root space and soil, sufficient light and nutrients and adequate water – plants will grow. A well-landscaped zoo will be far more welcoming and will present the animals in a much more natural way. Hopefully, it will also build awareness of the need for the conservation of habitats for wildlife. There is much evidence that many people feel more content, work harder and better as well as recover from illness quicker in a well-planted environment. Zoo visitor numbers annually around the world are at least 700 million. This gives enormous potential to showcase habitats and the need for plant and animal education and conservation. Plus, visitors will appreciate the care taken to keep the animals in a more realistic setting, giving them a much more stimulating daily experience. It is a win-win situation for both the animals and the visitors and much more interesting work for the horticulturalists.

Linked to planting, there can often be specific educational opportunities. Careful plant choice can result in many, many stories to catch the attention of visitors. This gives educators a chance to talk or use display material to help make visitors more fully realise the importance of conserving habitats around the world and even encourage them to take decisions in their daily life that reduce their impact on the natural world.

As I looked at the various problems I encountered, it was obvious that some plants could cope with very drastic animal damage, but often problems showed potential solutions too. 150 geese with access to entire lake edges and lawn areas caused a horrific mess from their droppings and also grazed the grass far too tight. The grass suffered, but many weeds growing in the grass did not – the geese were selecting only the grass to eat. They actually worked like selective grass killers, constantly keeping the grass down, yet allowing daisies *Bellis perennis* to grow well, for instance. The daisies were given copious doses of goose droppings to encourage growth. It's true, geese do not eat everything...

Retro fitting plants into existing areas, whether in a zoo or not, is always more challenging and expensive to do. It is essential that planting is considered from the first idea in master-planning, through the detailed planning stages to the actual building itself. Plus, there must be a team effort from all sides. The designers, the architects and engineers, the horticulturalists, the builders (not just the site foreman), the actual animal team and particularly the machinery drivers must be involved from the beginning of the project – nothing causes as much damage as quickly as an uninformed excavator driver!

As part of the design process, it is necessary to get everyone's ideas of what they believe the appearance of the area should look like – this should be part of the masterplan. Once the desired appearance is defined, then the critical and interesting task of plant selection and the thinking through of all ideas are the next important stages. They are usually highly enjoyable too. At this stage, there is the crucial point of instructing the architects to allow for soil areas for future planting and for these instructions to be given to the builders to get them in place. Often this is a challenge and it may need explaining more than once. Topsoil is the plants' foundation and if it is too weak, the planting will fail.

Hopefully, these notes will assist fellow horticulturalists to green up their zoos and dare I say, may even be of interest to a more general audience. People learn by looking at the complexities and intricacies of trying to grow plants in a largely animal-centred garden. The chapters follow my lessons from the field. They evolved as the zoo evolved from being a Victorian dinosaur to a modern model and I became ever sneakier in how I kept up with the work – hopefully while still practising good horticulture in more efficient ways. I particularly hated seeing work being done twice due to lack of forethought or work being generated or made awkward by poor design.

CHAPTER 1

Early Days

A Victorian Appearance

In the early 1980s, the zoo had one large lake, a smaller separated lake (but the same water body) for small waterfowl, lots of tarmac pathways and flights of steps and very few plants, mainly trees. Many of which were in poor condition, often due to soil compaction and root issues from too much tarmac. The lake had no planting along the edges and no fence. Between the geese eating everything and visitors walking to the edges, there was a very bare expanse of what was termed soil, though by definition much of it did not qualify as such. Many paths had a simple wire strand at 300mm high as a barrier and visitors were expected not to cross this.

Many animal areas were surrounded by a mown grass strip, maybe a metre wide. This is called the 'stand-off area' in zoo parlance. It is supposed to keep visitors back from the animal barrier – moat, fence or whatever barrier is in place. There was an element of the zoo being mini empires. Each animal team fiercely defended keeping their area exactly as they wanted it. Equally, the visitors had only one intention – to see the animals. If that meant climbing over the simple 300mm high wire strand, walking on the grass or any plants there, so be it. Gaining 150mm in height or getting 300mm closer was justification to climb up, on or over anything – wire, wood, metal, grass, plant, tree... And if they thought that they could attract the animals over by feeding them plant material, they would break a piece off or pull it up and throw it in because it might work – even if it had not for the last 350 people that tried.

As in any garden, there was always more work than time available. Reducing the amount of time spent cutting grass was an early first step. There was little point in much of the regular grass work that had been carried out. Let's say the grass needed cutting at the elephant/giraffe stand-off. This was a mix of grass and stretches of the common groundcover plant *Cotoneaster salicifolius* 'Repens' in patches one metre wide and a few metres long. Each grass area was separate. So, grab a wheelbarrow, edging shears, brush, shovel and a hover mower. This was the only mower suitable as it floats over the surface, over the edges and gets under the edge of nearby bushes. Get the mower started (often a task in itself), mow the grass (3mins) while blowing the bits all over the place… Cut the grass edges with the edging shears (10mins). Sweep it up and shovel into the barrow (15mins). Put mower in barrow and move on 5m to the next little patch of grass and repeat the procedure. Let's say it took thirty minutes per patch, times twelve patches – that is six hours of work a day gone. Then the next area… Half a week lost every week cutting stupid fiddly grass. It all needed doing again in the next week or so. None of it looked good, so there was very little job satisfaction. Plus, much of it had to be done before visitors arrived and the mowers kept breaking down…

The lake itself was a central feature and a real asset as it meant visitors always had a view across it, making the zoo seem more open. Its appearance was very worn down, with eroded edges to the banks and odd rubble and stones visible in the water. It had been used as a dumping ground for a great variety of material from old parrot cages to small vehicles over many years. Geese ruled the entire bank by overgrazing what little grass there was, bringing out enough water to make the edges muddy and leaving heaps of droppings all over the place. The few trees or shrubs that were there, were suffering from the associated soil compaction. Even small geese will cause this by constantly walking on wet soil. This was a problem that affected a lot of zoo areas. With no vegetation along the banks and geese climbing in and out wherever they wished, the edges were not only bare, but were eroding from the water constantly lapping against them and the action of the geese. Though there were a couple of areas where plants did grow – meaning it was possible.

The lake water level was reliant on rain. Heavy rain meant the level went up and sometimes we had to open a penstock (small sluice gate mechanism) to release some water which could flood further downstream through The People's Gardens in the Phoenix Park before entering the River Liffey. Equally, dry weather in summer could lower the lake and there was literally nothing that could be done about this. The lakes were held back with an old Victorian soil dam, no doubt with a clay core in the centre, but probably with a few leaks.

Each summer the custom was to clean the paths and lawns with a fire hose. A novel idea and incredibly effective in the areas covered. Dust and debris built up daily. All road cleaning was done by hand, with a good old yard brush and shovel. It was very slow, though effective along the edges, but often left quite a bit in the middle. It would be impossible to sweep the entire road width by hand. The fire hose certainly worked and it was the only way to remove the goose droppings from anywhere, especially the dry lawns. First wash off the lawns, then around the restaurant area, Roberts House and down towards the front gate. But water has its own ideas... A gentle torrent of water and debris, including an excess of cigarette ends, made its way downstream. It leapt down the first flight of steps, helped by extra water from the staff who knew this would happen. Bucket loads of dirt were washed down. On the next steps the same thing happened. It was all washed downhill towards the front gate where it turned to the lake. Ah, that's where it ends up! The lakeside road was done last and the dirt was slowly pushed into the lake. Did it sink? Not at all. At first it floated out as a thick skin. It slowly absorbed water overnight and sank over a day or so and maybe even travelling down towards the penstock enjoying the scenery as it went...

Planting up the lake edges revealed the history of dumping there. Ground had been reclaimed slowly from the lake every time there had been building work done. Any dry filling arising was tipped into whatever area was being worked on. This meant there was plenty of old concrete, bricks, steel bars and whatever else various builders had thrown out, topped with a thin layer of poor soil. A real journey of discovery but not quite archaeology in every spade-full. That explained the awful appearance along the eroded edges of odd bricks and blocks, steel bars, old netting, etc...

Lake edges about 1986, after initial planting of *Carex pendula*. Note the gaps between clumps after some losses, the uneven lake edge due to erosion and the wee green fence that was used to keep geese off.

Lake edges during planting. New erosion barriers of timber or stone, lots of short plastic goose fence and a few geese observers showing interest. Paved sections allowed access for the geese, chosen at the shallowest points for safety.

As spring started and days lengthened, there was a warming of the lake water and a gentle increase in algal growth. The main species was a blanket weed. This formed mats on the bottom of the lake. During photosynthesis they released oxygen, which sometimes got caught up in the mat of algae itself, unable to rise to the surface. Then it became a balancing act. Eventually small pieces, no more than 100mm wide, would float to the surface. As sunlight strengthened, this became more pieces, until you could see bits floating all over the lake. If there was any flow through the lake system, the mats would collect at the penstock.

Can we clean the lake? This would be tricky because each mat is made up of tiny fibres of the algae. Touch it and it falls apart, sinks to the bottom again, only to rise later. There was no stopping it.

Why so much algae? Well, the lake was very well fertilised. Imagine the quantity of fertiliser from the geese that were fed a sack of grain per day and the flamingos too. Excess food in the water could not be removed. At that time, most of the buildings around the lake drained into the lake – the rhinos, the orangutans, the gorillas, the brown bear and (let it be whispered) even the lakeside coffee shop. On its own, it turned one end of the lake white when a freezer broke down and sixty litres of ice cream went down the drain one morning. No wonder the

Peafowl and guineafowl scratch soil and neatly laid mulch looking for insects. This will ruin a nicely planted area, scattering debris everywhere – over plants and over the path and exposing the soil to light again, so more weeds and water loss. They also go around mob handed, two or three females and as many chicks as they have (always too many in the gardens teams' humble opinion). But some plants were not eaten, including anything with a scent. Meaning shrubby *Salvia microphylla* was safe for instance, as was some of the *Cistus*, though that was the sticky foliage. The summer bedding was another target. The 'normal' mix of bedding *Lobelia* and *Salvia* would see the peafowl eat every single first flower head off the bedding *Salvia*. The next year marigolds were tried; strongly scented, so they were left alone. Praise be. Though it did not stop them or the jackdaws *Coloeus monedula* pulling up the plants just after planting to get at any insect, especially worms, underneath! Eventually netting was put up for the first month.

High season (June, July and August) simply became too busy after mid-morning to do much on the visitor side. Routine was to attack the boundary hedges once it became crowded, maybe a little mowing or deadheading work before the 10am break, then onto hedges. Most were the simple cherry laurel *Prunus laurocerasus*, but there was a lot of beech *Fagus sylvatica*, a much nicer hedge, but not as forgiving if pruned hard. Though some were monsters, the largest beech was only 100m long, but three metres tall and worse, two metres wide. Two rows of young plants had been planted thirty years previously, but instead of being half a metre apart in the row and rows being no more than 200mm apart, these rows were more than a metre apart to start with, hence the width. It was difficult to reach across from the top of old wooden step ladders to cut the top centre. Something had to go. But beech hates hard pruning. After carefully checking, almost all the back row of beech was cut out, taking care not to cut off any branches that had managed to get to the other side. It meant a lot of three-metre-tall beech with a dense clump of branches was being cut down, the amount of arisings was going to be significant. That was until someone pointed out that the elephants nearby loved beech as browse. Every day or so, one or more beech were cut down and dragged the short distance to the elephant house. The elephants milled into them as they got very little natural browse at that time and left only the stump of about a metre long. This saved us an awful lot of work. Four-legged tree chippers! Recycling at its best and less work for the gardens team or waste disposal.

Around other parts of the boundary hedges, it was just plain awkward to get into or out of. Sometimes we needed short ladders to access the top of the hedges. Initially, only electric hedge trimmers were used or hand pruning with secateurs and loppers. There had been a dearth of tools for the gardens team for years. Getting new secateurs as personal equipment for each gardener paid off here. One Felco secateurs and you are set up for life. The trickiest problem though was getting rid of the arisings, especially large quantities in narrow areas, which if dry would have been quite a fire risk. Removing them worked but was very slow. They had to be put into a skip – time and money. Some areas were really awkward, until there was a suggestion to bury them... That suggestion was met with 'How?' from the team, as there was so little room to work, between a wire fence and the roots of the hedge itself. After lunch, out came the mattock – a good old-fashioned tool that's great for hard ground or cutting through roots – not that any real harm was intended to the hedge. A trench 250mm wide, only 200mm deep and as long as was needed, was teased out with much swearing. Then drag the pruned off hedging into it, *Griselinia littoralis* and cherry laurel *Prunus laurocerasus*. We filled it in two minutes. Then Tom, with his weight advantage, walked on it. It sank. More went in again. Another walk please, Tom. It sank again, but less so. Another layer, another walk. The last layer, but we were now about 250mm above soil level. We threw the excavated soil on top, walked on it again. Hey presto! With that little weight on top, it stayed down too. Though you could certainly see where the trench was. Turn and walk away. Don't even look at it until the next hedge cutting session a year later. Lo and behold, all that was left was a slight mound, which was much easier to dig out the following year, with almost nothing left of the hedge material. Worked well for several years until the whole area was redeveloped. Nowadays there are people praising 'hügelkultur,' a technique used by gardeners practising permaculture for making raised beds with a lot of woody material in the centre, for vegetable growing, we just had a very, very thin version of it.

A lot of this work was in summer and behind the scenes when too many visitors were around for general work. In places it meant cutting back overgrown hedges, which made it bare for a while. Good hedges really need annual pruning. This had not happened as animal teams often wanted no visibility and sometimes complained if work was done near them, but the best hedges are kept thick by regular, preferably annual attention, not neglect. It only remains to say that it's amazing how much hedging you can get done when the animal team staff take two weeks leave in the summer. Besides, it's very hard to stick it back on...

CHAPTER 3

First Small Projects

Visitors often asked the names of the plants and for a while, simple laminated paper labels were used on any plants that had a story. These gave the origin of the plant, if it was a cultivar, where it was found, a little note about it and maybe the family name too, with a little note about that. Eventually, it was another stick to beat the team with, especially replacing old labels, so it stopped. Though writing the labels, checking the notes and reading up on the plants was very educational for everyone. But it did instigate some, umm… interesting conversations. For instance, a sign on a Lombardy poplar *Populus nigra* 'Italica.' This is a fastigiate tree, with a very narrow habit. The original tree was found in Lombardy. Propagated easily by cuttings, it found its way into the rest of Europe in the early 1800s. It is a male clone, no female flowers are produced and it possibly arose on an ordinary Black Italian Poplar as a sport – a botanical term for a variation in the normal growth pattern. Gardeners will often see a different flower colour or leaf variegation – all these are 'sports.' It is a member of the willow family. So, the label on the tree was something like this:

Lombardy Poplar

Populus nigra 'Italica'

A male sport found in Lombardy, Italy in the early 1800s

Salicaceae – the Willow Family

That's all fine until one day a gentleman visitor came with a question. He had to ask three times before the question made sense. He knew how to play football, cricket, rugby, hurling, et al and wanted to know all about this game called Lombardy poplar and how to play it. He was right to ask, you can read it both ways, so a valuable lesson was learned…

Another time the spelling was challenged. A double form of the gean *Prunus avium* 'Plena,' a wonderful plant when in full bloom with its full white flowers, labelled thus:

Double Gean

Prunus avium 'Plena'

A double form of the Wild Cherry or Gean. Believed to have been found in France early in the 19th century.

Rosaceae – the rose family, which includes many of our fruit trees; apple, pear and plum.

Taken to task here for spelling 'gean' incorrectly – surely, it should be 'gene' if the genes are doubled?? Again, a valuable lesson learned. A story much later from an educator in a UK botanic garden that you should never underestimate the ignorance of your audience... And that was after a tour where they stopped to look at *Camellia sinensis* – the plant that gives us tea. Looking at the typical *Camellia* leaf, 75mm long by 35mm wide, someone who really should have known better said, 'I thought tea leaves were much smaller'...

Feelings were a little mollified when other signs were seen, especially when temporary signs went up in the South American House after refurbishment. They were written in extreme haste and not proofread by any technical staff. The actual text was poor, with many typos, capitals where they should not have been, etc. Then you read that the two-*toad* sloth was sharing its habitat with the iguana... Can you see them cohabiting? The plant signs did not seem so bad!

The Reptile House was rebuilt by a community employment scheme, reimagined as an East African Reptile House. This was the first redesign of an existing house. It was an interesting process but very slow as it was done with trainees. So many hands, but often in the way of each other. An interesting lack of experience led to minor errors, such as building it with access doors wide enough for a wheelbarrow with an immediate, sharp 90° turn. You could get a wheelbarrow through, but only if carried vertically. The point refused to be accepted that the whole idea of getting a barrow through was to have it full. Planting was discussed. It was decided it would be good to try and use plants from East Africa, and some from South Africa. Plant selection was easy, as many house plants and garden plants come originally from here. This decision was mentioned to our council members and they requested a list, which was something like:

Indoor Plants – *Tradescantia, Sansevieria, Zamioculcas, Ceropegia, Kalanchoe, Saintpaulia*... and many more.

Outdoor Plants – *Sparrmannia*, *Hesperantha* (*Schizostylis*), *Agapanthus*, *Euryops* and many more.

That was grand, until the Director realised that he would have to try and pronounce these names in front of the zoo council… Common names were then requested. Ok.

Indoor Plants – wandering Jew (*Tradescantia*), mother-in-laws tongue (*Sansevieria*)…

Outdoor Plants – African hemp (*Sparrmannia*), Kaffir Lily (*Hesperantha*)…

Old style, not politically correct names – most not used now. There was no way he was willing to say those either. Drop that list quietly please but use the plants.

The *Sparrmannia africana* has remained a favourite plant. It is not hardy and needs a greenhouse to overwinter but grows easily from cuttings in summer that will give two-metre plants the next year, with good large leaves to about 150mm across. The white flowers are 15mm across and have distinct stamens of red and yellow, which are haptonastic – otherwise known as thigmonastic – they move very quickly outward when touched, to aid pollination. Colleagues in another garden mentioned that during a tour when the plant was pointed out, a very refined lady at the back of the group called out, '*Sparrmannia*? We used to smoke that in Kenya.'

Unfortunately, it proved impossible to grow plants well inside the new reptile house, there was no natural light and any grow lamps were too far away. There was also the issue that to save money only well water had been used. The zoo site is on limestone and the water had a high pH. Water hyacinth *Eichhornia crassipes* – a real thug when happy, clogging many miles of waterways around the world – disappeared under the water in the first two days, never to be seen or tried again. Many of the other outdoor plants have since been used elsewhere in the zoo to great effect.

Some other little projects were started too, tackling the worst areas. There were still signs of the old zoo around. One was an old-style bear pit. This was literally a hole in the ground with three-metre high concrete sides that visitors used to peer into. Fortunately, that had already been modified and used for many years for prairie dogs *Cynomys*, a burrowing rodent. It never looked good. How could it with poor grass close-cropped by the animals and many bare areas as they dragged the grass roots in as bedding. This was converted into two small exhibits – one for otters and one for beavers. This was a totally new way of doing things. Staff were asked for ideas and comments and helped to design the area. It was also a great lesson in how much control you had to have over the build. The old walls and floor from the bear pit would certainly keep the otters and beavers in, a simple double chain link fence was erected along the middle, with much simple tree planting, mainly common birch *Betula pendula* between the fences. The beavers would obviously cut down any trees in their area and this screen would separate them from the otters and give a better, greener wooded appearance.

The beavers had a wee stream, but the builder could not think of any way other than a wide dish shaped slope. He was most upset when we put in additional rocks to make it a lot more natural. A simple sump collected the water at the base for recycling with a pump. All worked well and there was even hope for a little dam-making. The otter also had a smaller water feature – a wee waterfall over a large slate slab. After a few days, the otter was nowhere to be seen. A few naysayers were quick to point out where it got out, you can see the hairs on the wire. Told you it wouldn't work... A good few days later, there was movement. The otter had been hiding under the waterfall rather than in the house provided. 'Ahh, it must have fallen back in,' was the comment... Typical.

The project taught the team a lot about design, the importance of involving everyone, that all ideas are worth listening to, even if not possible, but most important, it showed how close the gardeners would have to work with the designers to be able to get plants in and with the builders to be allowed to do it at the right time.

One year a dinosaur exhibit called Dino-Live was planned, obviously using models. That fact did not stop some people phoning up wondering what they were fed though... The life-size models were commercially available to rent and they moved with a silent pneumatic system. A large marquee was put across the main lawn to accommodate them. The design was fixed between the educators and the horticulture team and a list was drawn up of what plants would look 'Jurassic' to show off the dinosaurs best. The Jurassic was from about 215mya (million years ago) to 145mya, when the supercontinent Pangaea broke up and life recovered remarkably after a mass extinction. Gymnosperms – naked seed plants such as conifers and *Ginkgo* – became dominant, taking over from ferns. Easy enough to say the planting should aim for conifers and ferns, but what was available and in what sizes? The design was simple, a path around three inside edges of the marquee, with a two-metre mound of soil in the centre to stop people from seeing each other, all densely planted. Root-balled 3-4m willows and poplars were used (no big colourful flowers) along the very centre of the mound for height, with conifers lower on the slope, mainly Norway spruce *Picea abies* as readily available at 2-3m, with a few nicer specimens of deodar *Cedrus deodara* nearer to the paths.

A couple of maidenhair tree *Ginkgo biloba* had to go in as they had such a good educational handle. *Ginkgo* is often found as a fossil, almost identical to modern-day leaves. There is a small fossil leaf in the Discovery and Learning Centre. This remarkable tree, now only found naturally in two places in China, is the last surviving species; fossils show more variety. It is long-lived with wonderful autumn leaves of glorious yellow. The trees are dioecious, male and female flowers on separate trees. Pollination technique resembles cycads and fern. The resulting seeds have a disgustingly foetid smell, which means most trees planted are male, though they have high allergy pollen. Sometimes you cannot win. *Ginkgo* has many interesting toxins, often with antibacterial or medicinal uses, which arose as useful defence toxins. These include trying to stop animals browsing and if you think rabbits and deer are a nuisance, imagine a ten-tonne *Diplodocus* nibbling away at your garden *Ginkgo* feature…

Loaned to us from Trinity College Botanic Garden, a large and quite old cycad *Cycas revoluta* was another great highlight. It was carefully positioned under a twelve-tonne browsing *Pachycephalosauros*. *Cycas revoluta* is native to the southern islands of Japan where sago is made from the starch found in its stems. Washing out the starch thoroughly has to be done very carefully as cycads are extremely poisonous, possibly to reduce animal browsing originally.

There was a tall screen along the centre and some interesting Jurassic age plants, but something dramatic and in bulk that caught your eye was needed nearer the visitors. Many Japanese aralia *Fatsia japonica* were used. These are totally hardy, but if grown outdoors they can look a little weather beaten. Very cheap greenhouse-grown plants, which arrived in five-litre pots at almost a metre tall, were used and really made a difference. It is a member of the ivy family *Araliaceae* with very large palmate leaves, a deep glossy green, up to 300mm. This is especially so when greenhouse-grown, as it is also sold as a house plant. The flowers are borne in late autumn and look just like common ivy *Hedera helix*, creamy-white, but in much larger umbels up to 200mm across, an excellent nectar supply for insects and very useful for late autumn feeders. As an understorey, loads of ferns went in, but commercially these are quite small when sold and never really filled out the spaces in the one summer. A small pond was put in, originally a wee plastic mould that had been donated, but it was far too small and ornamental. Praise be, other people thought the same when they saw it and moaned, ensuring it was replaced with a simple plastic-lined odd-shaped pool. The retaining sides were made of the grass sods removed to make the necessary hole and the surrounding sandstone rocks hid everything nicely.

It was such an unusual project and with such unusual animals, there was bound to be a few memorable moments.

Once the marquee was up, tonnes of sandstone rocks were delivered to delineate the paths and when the main topsoil was in position, the dinosaurs arrived. Three large lorries arrived with several mostly fully-assembled dinosaurs. As they were lifted off the lorries with a forklift, they had to be moved, but where to… Even getting into the marquee would be slow and tricky. The drivers needed to unload quickly and get the lorries out.

Ok. Park the dinosaurs at the ring-tailed lemurs *Lemur catta*. There was room enough, it was all on tarmac and near the marquee. The first three came along, all plant-eating species. The lemurs came over to the fence, climbing up and down looking most perplexed and interested. Then along came a half-life-size *Tyrannosaurus rex* with a wide-open mouth showing plenty of bite. The lemurs went ballistic, climbing up and chattering away. No matter that they had never seen one before or that they evolved millions of years after T. rex died out, they recognised a carnivore and were trying to scare it off!

We had also been loaned a fossil *Ichthyosaur*, which looked a little like a small dolphin. For the opening night someone slipped a small label into the glass case stating, *Caught in Durrow, Co Laois, May 23rd, 100 million years BC*. An obvious nod to a well-known fisherman on the staff.

As the project developed, curious eyes were trying to see what was happening – hard to hide seven-metre-long models and impossible to have all doors shut all the time. Other staff would pop in sometimes and occasionally brought their children for a sneak preview. Each animal was connected to power and pneumatic lines with a control box for each hidden behind it. One day a couple of kids came in with their father for a quick look. While standing chatting, someone moved their foot nearer to the power button. As the kids all looked away, the start button for the pneumatic air supply was gently tapped by a foot. When they looked back, the mother *Diplodocus* head on its three-metre neck was swinging towards them… Screams galore ensued as they were not aware that they could move at all!

Once the exhibition closed, there was the opposite job, stripping it down – never as much fun as building. While walking into the marquee one morning, some staff were discussing how to drain the pond. They were surprised when they got there to find no water. Some sneaky character had thrown a garden fork into it the afternoon before, allowing gravity to do its work overnight.

All the plants were reused in the zoo. The poplars and willows went into the otter and beaver area mentioned above. The *Fatsia japonica* went all over the zoo and have formed some fine specimens. There had been only one in the zoo for many years, located at the back door of the bat house and you walked under it as you came out. It had been said to have been planted by Robert Lloyd Praeger, the renowned Irish naturalist, who had been on the zoo council for many years. He probably put in other interesting plants too, such as the box leaf azara *Azara microphylla* that confuses people with its vanilla/chocolate aroma when flowering in early spring, but the flowers are so small no one notices them and a tree aster *Olearia virgata* var. *lineata*, a small-leaved plant used later in the savanna planting – see below, but all were grown from these original plants.

The entire project had been designed by the education and horticulture team and fully built by the zoos in-house garden and maintenance team. A remarkable achievement for the small team and they enjoyed it as well. The project did well too, especially as the film Jurassic Park came out the same year. It seemed to change how people thought about the gardens team, especially how plants could be used in a much more interesting way. The 'old zoo' mentality was slowly changing.

For many years, the zoo had been held back due to lack of funds. It was totally dependent on income from visitors. In a vicious circle, too few visitors meant not enough income to develop, so less visitors came. Once granted a moderate development fund from the government, the most urgent needs could be considered. These needed to be a balance of visitor needs, animal needs and general infrastructure. To reduce the waste entering the lake, one of the very first projects was a new main sewer system, very unglamorous although interesting when they laid the pipes. All existing drains from buildings were tapped into this as they went. Unfortunately, there must have been a little confusion at points and a few pipes that led into the lake were tapped back into the drain too. This only became apparent once the system was up and running and the lake level and flow rose with rainfall.

The gardens team looked after the lake, including watching the level. They wondered why there was no overflow from the lake despite a very wet winter and started asking awkward and unwelcome questions. While the orangutan outdoor area was being enlarged and improved, the lake was dammed to allow work. Looking at a little trickle of water coming through the dam, the question was asked, 'Where is that wee stream going?' Into the lake was the answer. It was pointed out that the lake surface is one metre uphill from the wee stream – very clever water to beat gravity. Let's just say the pumps for the drainage system were both working flat out to keep up… Much camera work checking the drains was carried out and a good few 100mm 'leaks' from the lake were fixed without comment, just a few sideways glances when the gardeners went past.

New islands for primates were also created, giving much better conditions for several primate species. These included a larger outdoor area with access to a heated building on the shore if desired, more natural climbing frames of timber and rope and a little more naturalistic for the visitor too. The lake is man-made and about 150 years old. The bottom of the lake is not solid, but a varying depth of fine silt, mud and organic debris – think of all the tree leaves – which defies description but may best be described as sludge. Stone filling created new islands of about 10m square and maybe two or three metres deep, from solid lake bottom to new soil level – a cubic volume of say 250m^3 or at least 250-tonnes. When you tip that much weight onto a sludgy bottom, it does not necessarily bury it, it can push it away instead or cause a wee underwater sludge wave. In places, this displaced the water so much a boat could hardly be rowed around some edges. Any sludge that is buried or mixes in among the stones, will slowly settle further. Sure enough, the new islands slowly sank a little over the next few years.

The other problem this sludgy bottom caused was that when made, each island was surrounded by a rock armour of large stones (250/500 kgs) that was meant to stop the edge from eroding. Over a few years, these slowly sank 250mm or so, often leaving gaps between them where water could get in and erode around them, making them sink even more. Eventually, as always, the gardens team were asked to do something. Yet again, pendulous sedge *Carex pendula* came to the rescue in large mature clumps from elsewhere in the gardens, along with lots of yellow flag iris *Iris pseudoacorus* – again large clumps were used. The only limit on size was what could be managed in the boat used to row out and back. The *Carex* worked but was a challenge at times as the geese tried to graze it and the swans tried to pull nesting material out of it, so some became dislodged despite having timber stakes driven into the clumps. The iris was really good. It was planted in winter when there was no growth – the rhizomes grow in all directions and overlap each other. The feeding roots grow down through all this, making a really solid mass and as the leaves are unpalatable, they were not touched. Once established, it spread out slowly ensuring the water erosion ceased. Some clumps that were well attached to the island became nest sites for the waterfowl, even mute swans *Cygnus olor* nested, happily sharing the island with ring-tailed lemurs *Lemur catta*. *Iris pseudoacorus* is a great land-builder, a strong performer in the dynamic and perpetual battle between land and water as the fine roots trap silt and the rhizomes make a solid tangled mass, fit for other plants to root into.

Ring-tailed lemur island. Climbing structure very obviously not natural. Note rock armour in places sinking too much. Plants in gaps include *Caltha polypetala*, *Carex pendula* and *Iris pseudoacorus*. At the rear are very old plants of *Phormium tenax*. The dead leaves are from severe frost damage three years before; even dead and frosted, they will not rot. The lemurs were gracious enough to share the island with a nesting mute swan.

There is one big drawback with islands, how do staff get on to them to work? When the first islands went in as part of the development, there was a habit of using a simple rowing boat and that's grand. Then all the gear needs to be ferried out – tools, plants, etc. With two staff on an island, if one had to row back for something forgotten earlier, perhaps an hour after the morning tea break, it would leave the other gardener marooned. 'That's handy, I need to nip to the loo anyway,' says the rower. Funny how the mind works. The poor lad marooned on the island then got to thinking and in twenty minutes he was in extremis... The only cover for his predicament was a long log about two foot thick. Kneeling down behind the log, facing a busy visitor path twelve metres away, he managed to make himself a lot more comfortable, but his knees nearly got wet. Later on, we got into the habit of hiring aluminium access ramps as temporary bridges. Much handier and more, shall we say, convenient.

The islands would get very little follow up horticulture work once the primates were there. Planting progressed, the plants needed a good watering in, enough to really soak the soil. There were no taps on the island, so a hosepipe was rowed across. Now that's handy enough, though the only available tap was on the furthest bank away. Full of air, the hosepipe actually floated, but it has a natural recoil when brand new and just unpacked – rowing it across was akin to rowing against an elastic rope, but we got it there. Mulching was another necessity as it would reduce water loss and drastically reduce weed problems. The mulch was wood chips from tree surgeons. Very handy and cheap. One staff on tractor, one in the boat, two more on the island and maybe eight or so 75L pots of mulch in each trip. Worked well, but so many miles were covered rowing. And no one fell in on that shift!

The original designs for the islands had quite a lot of planting, but this was thought to be excessive for the very reduced team in the gardens at that time. Contractors came in, but they were told to run planting schemes past the gardens team first. The initial design had loads of colour – variegated plants with white, yellow or red leaves – imagine any of the primates against a backdrop of variegated *Cornus*, *Sambucus*, *Weigela*, etc. It would have been awful. The plants are the background within the habitat. This planting would have drawn your eye to the plants far too much. Hopefully, the gardens team is forgiven by now for saying 'No, no, no.' to almost every plant picked. There were only four criteria: plants had to be green, very tough, be able to be pruned hard if needs be and not be poisonous. There was no attempt to plant animal-resistant plants, though probably not possible then, what with the combination of primates and geese. Fences that keep geese out would be pulled up by primates. Fences that stopped primates would not stop geese. The poor landscapers were aghast to come in on one Monday to find that the geese had eaten most of the previous week's work, as they had not put back the temporary fence. Equally, many of the plants were edible, especially to some of the primates that are mainly leaf eaters in their native habitat. They can even tackle some poisonous plants.

To prevent any of the primates eating the planting at each end of each island, electric fences were put in. Expecting the primates to climb, they had alternate earth and live wires, so even off the ground a shock was possible. They didn't stop geese – either the feathers insulate or they do not register the pulse, whatever, they pushed in and demolished a lot. Each primate species had different ways of getting past the fence. Lar gibbons *Hylobates lar* and spider monkeys *Ateles fusciceps* climbed with natural agility. They could sense which wires were live and carefully grasped the stakes, climbed up, swung over and climbed down within a day or two of going on the island. The colobus monkeys *Colobus guereza* had the most amazing leap, from a sitting position they seemed to clear the two-metre-high fence easily.

Irrigation had been installed once mains water was plumbed to the houses on the shore, simple rotating sprinklers for some of the islands and if it was noticed that the plants were being attacked, the water would quietly be put on. Suddenly the plants were left alone, until we turned it off. The primates have great eyesight and they are astute enough too, so after a few days of this water on/water off treatment, they learned to get out when they saw any of the gardens team approaching and no water was needed. Either way, most plants did not last long. New Zealand flax *Phormium tenax*, was about the only survivor. It was just too tough, but it caused a problem after a few years – if young colobus monkeys jumped into a clump from a high point, they could not get out easily as they could not get a grip on the leaves.

The electric fences had issues as well. The builders thought they could just push the plastic posts in the same way you would a timber post, using a small bucket on a mini digger, but the plastic bent in the middle, until – Thhwock! – it snapped and two or three ballistic bits of post flew off! The foreman's language was choice as one piece had gone past his head with a noise like a dying partridge. Holes needed to be dug instead. The fences kept shorting out, which turned out to be because they were made of recycled plastic. They had small air pockets within them, which slowly filled up with rainwater, the wires went through the post and then if you had one live and one earth wire contacted by water, a short happened again and again. Apart from all that, they looked really awful too.

After the islands, the next project was several new areas for cats, tigers, snow leopards and jaguars. Designs were drawn up. They followed the old school method of pushing the animal area as tight as possible to the next fence line, leaving very little room for planting. Sides and boundary were simple laurel; easily purchased as one-metre plants, guaranteed to do well and give a green background all year. At the front, in the stand-off area, it was very different. Screens were requested for all the fences, only allowing viewing at the designated glass panels. Not a problem. What budget is there for plants? A pitiful amount was uttered to plant the three areas. After much discussion, this was increased fourfold, especially once it was realised that there was actually a reasonable horticulture budget there, but it had been held back as a contingency for 'any other works.' Add in the topsoil costs and the new budget was about right. A lesson learned, certainly by the gardens team.

It was very limiting with only a narrow 1.5m width stand-off between the animal fence and the visitor barrier. The jaguar *Panthera onca* needed a full dense screen, so a simple (and cheap) laurel hedge went in here too. The snow leopards *Panthera uncia* had no room for a stand-off; a timber fence hid all between the viewing areas. However, there was room nearer the house where a simple planting of white stemmed birch *Betula utilis* subsp. *jacquemontii* was chosen, underplanted with Christmas box *Sarcococca confusa*. There was scope for a little bit of theming too, as both these plants come from the Himalayan region. It was good to see plants and animals from the same area. The same birch was also planted in the snow leopard habitat, but they needed protection with electric fences to avoid the cats scratching the bark – typical behaviour for any cat.

Normal planting would have spaced the trees maybe three or four metres apart, but to make it look more natural, they were planted with two or three trees very near each other or even in one spot. This creates a much more natural, less gardened effect, especially if the trees are planted at different sizes. Practically, it makes it easier to run power cables to protect the trees with electric fences. A few *Carex pendula* crept in here quietly. They broke up the landscape of limestone blocks nicely, but they were so happy they seeded everywhere! Years later, there were dozens of good thick clumps. The snow leopards loved them as they could lie down almost anywhere and disappear from view...

The tigers could really have something interesting now there was a budget. Why do tigers have stripes? For camouflage, to disguise their body shape when stalking prey. Which Asian plant has good strong straight vertical lines to match the stripes? Bamboo of course! With a few dozen plants in 12L pots, within two years there was the perfect stand-off planting that hid the animals almost completely, unless you looked really closely. Part of the new zoo design was making it a journey of discovery for visitors. The more you looked, the more you saw. The first design here did no planting within the habitat – more on that later.

Bamboo is a most interesting plant. Many different sorts from Asia, North and South America and Africa. Used by gardeners for many years as focal points, ground cover, quick vertical screens, interesting stem features, a wonderful selection. It has features that make it unique. Firstly, it is a grass and can grow very quickly – very, very quickly. The bamboo stem, officially called a culm, is almost always hollow. At intervals there is a bump, officially called a node, which is solid. At ground level, look for a new shoot only say 250mm tall. If you can spare it, cut it off and split it down the middle. Look carefully and you can see the hollow culm, with the solid nodes at intervals. When the culm grows, it doesn't just grow like trees or shrubs at the end of each shoot, it grows at each node. Count the nodes, there could be twenty or more. Imagine each node growing only 1mm per day. The culm will grow at 20mm or more per day and that would be quite normal for many of the species grown in northern Europe.

Tropical bamboos, with an eventual height of 30m, with heat and rain, can grow up to 1m per day. But this rate of growth has a problem – the stem is initially too weak to support itself. The plants remedy for this is the culm sheath. Look carefully and the new stem is surrounded by a thin but stiff outer cover that wraps around the culm. This is a temporary growth that falls off later. It is full of silica to strengthen it and the silica has another advantage, it makes the culm quite nasty to touch or eat, thus preventing a lot of animal and insect damage at a very vulnerable time for the plant, when it is young and nutritious. Some species have silica needles that are bad to grasp. Gardeners need to take care – it is similar to having lots of splinters in your skin, but they are like glass, hard to see and break easy. Not nice. The culm sheath eventually falls off and rots very slowly (all that silica) and makes the best fertiliser for the bamboo again as the silica can be reabsorbed by the roots. Silica in plants is a very interesting educational topic in itself.

Bamboo has one very significant feature. It is a grass, it flowers. All grasses flower you might think, but many bamboos flower only once in their lifetime and then they die. This is a strategy to wrongfoot any seed eaters in their native habitat. It is similar to there being mast years (a bumper crop) for oaks *Quercus* or beech *Fagus* in Europe. So much seed is produced that some of them will survive to grow and continue the species. But if one hundred bamboo of the same species are planted and it all flowers and dies, there will be a woeful gap for a while – and a hefty replanting job. To reduce the possibility simply choose two, three or more species and spread the risk of mass flowering. Of course, species is the important part. There are many bamboos that are selected cultivars – a nice habit or a good stem colour – within one species. Plant several cultivars of the same species and they will all flower the same year. The really fascinating aspect of bamboo flowering is that no matter where in the world they are, that species will flower the same year. No matter where in the world, north or south, east or west. No proof as to what the trigger is, each event maybe 50 to 100 years apart, depending on the species. Sunspots have been suggested or maybe somehow the plant can measure time – some species will flower after a very fixed number of years. Whatever it is, it has to be a global trigger.

Another area tackled was an old cheetah enclosure on the far side of the lake. This was largely a wooded area, mainly several evergreen oaks *Quercus ilex*, which date from 1904. After a violent storm in 1903 which caused a lot of damage including the loss of almost 3,000 mature trees within Phoenix Park, Lord Ardilaun donated 800 evergreen oaks to the park. At that time, this part of the zoo was still part of the Phoenix Park. Wolves were the chosen animal for the renovated habitat, but staff immediately felt sorry for whichever builder was chosen. The site was a steep slope, a narrow site and with winter fast approaching, the team could see a lot of mud and problems for them – and the trees, which form a great windbreak.

The trees survived despite small excavators, dumpers, trenches being dug for fence foundations, and, far worse – the need to have a two-metre-wide strip of steel mesh buried under the soil at the inside base of the fence to prevent the wolves digging out. Add all that up, plus traffic across the site and there was considerable root damage and compaction. Again, a steep learning curve of what building for an animal could entail. Evergreen oaks must be one of the toughest trees, as there seemed to be little reaction, some dieback was expected, branch or leaf loss, but they carried on. Other evergreen oaks within the zoo were equally tough with building work around them. Think about it... A Mediterranean species, drought-resistant, evergreen, the old leaves drop in summer when stress is at a maximum. It will grow back after horrific pruning. Severe coppicing seems not to kill them.

One very dry summer, one of the evergreen oaks dropped all its leaves except about ten on one twig. Other specimens in Dublin were as bad. They all came back later in the summer once rain resumed. Again, think about it – these are all seedlings. The acorns had been collected from somewhere around the Mediterranean. They are all unique genetically, some will keep leaves longer, some will drop quicker, but the important thing is that some of them will survive better. Natural selection. If you look at the wonderful groups of these oaks in the Phoenix Park (all the same age, from that 1904 gift) each tree is very different, leaf shape, tree shape and outline, leaf colour and degree of serration on the leaves – a wonderful example of diversity within a species.

The worst problem you must take into consideration when planting for wolves, is that they are very mobile and run around a lot. You may wonder why is that a problem and it's the same issue in many animal areas in many zoos. Constant foot traffic causes soil compaction, even with lightweight animals and especially when the soil is wet. The structure of the soil in the top layer – maybe only 15mm, but often much more – is damaged, air is pushed out and mud appears. Mud is literally a mix of water and soil particles. This layer, even if only very thin, reduces the passage of water through the soil, which makes it muddier and it rapidly gets worse. Air exchange is also severely affected and that affects the roots. The roots need air, which comes as a surprise to most people. Most vegetation disappeared very quickly within the new wolf area, though the tough evergreen oaks kept going, but soil compaction would become a regular issue with various animals around the zoo.

CHAPTER 4

More Developments. Bigger, With A Masterplan

The mid-1990s development carried on, fortunately with a second funding grant. The first had given us various projects, built largely in the old style with little room or scope for planting and with a gardens team that was severely understaffed in many ways. The second funding grant in the late 1990s came after radical changes in design philosophy. Out went pushing the fence as far as possible, out went flat areas, out went straight paths and out went leaving the plants as a last-minute thought.

In came a new masterplan, with external master planners that were not afraid to look at the bigger picture and think really big – though sometimes that needing restraining. In came day-long meetings where everyone was involved – the animal team, the designers, the engineers, but also the horticulturalists, the education team, the facilities team. Not all staff obviously, but plans were shared and discussed, feedback was welcomed and acted on. A very pleasant change. It was very useful and productive with comments from anyone about anything. This enabled everyone to have a better understanding of what was wanted and needed. Then the designers tried to figure out how to achieve it.

In came visitor needs – toilets, educational areas, play areas that were educational too and most importantly, carefully sited and angled viewing points – maybe with a roof for wet weather, framing the view, hiding animal houses and fencelines and other visitors. A very basic desire is for visitors not to see each other as it makes the visit more personal and more animal orientated. It restricted the amount of viewing by the visitors. This took a lot of getting used to by both staff and visitors as both groups were used to all-round constant viewing. This also had the reverse effect – if you cannot see the animals, they cannot see you! For many animals this is of great benefit. They are not under observation from all angles at all times. This helps satisfy some aspects of animal welfare.

The Five Freedoms of Animal Welfare:
1. Freedom from discomfort (an appropriate environment needed).
2. Freedom to express normal behaviour.
3. Freedom from fear or distress (if able to screen themselves from visitors or other animals).

4. Freedom from hunger and thirst.

5. Freedom from pain, injury or disease.

All of which add up to the guidelines for 'animal wellness.'

In came theming, taking inspiration from a specific area of the animals' natural range. This meant trying to make the new animal area look like its natural habitat with appropriate planting. There was discussion about keeping the planting limited to plants only from the animals' natural range. This had two major problems. Firstly, with the difference in climate, it could be very limiting. The second issue was the propensity for change in the zoo. If the animal changed habitat location, would there be a need to change a lot of plants? For each themed area, a plant list was drawn up so there was a basic planting palette for each area, not a detailed 'must have' but more a 'type of plant' and whether it looked right for that habitat. This was far more important than we realised, because if the same plants had been used everywhere, it would have looked the same. By having a very defined 'type of plant' palette, it became easier to make this work and give a varied look.

In came the visitor immersion experience. Visitors were meant to feel that they were entering the animals' habitat, on a journey of discovery and that could not happen on a normal pathway through the zoo. Narrower, more winding, informal, more personal trails were needed. This could not happen unless the pathways had more room to wander. The available area for the project may have only been 75% for the animal habitat, the rest for the visitor-side; that was a massive change in the basic design specification. Following on from that came the best development – in came so much more room for plants. This was a real game changer for the horticultural aspect of the zoo. Suddenly, instead of a 1m wide stand-off area within which 5m fences or buildings had to be hidden, there was now maybe a 10m wide stand-off area planting bed and maybe the same on the other side of the path, with plenty of soil depth, sometimes even mounding the soil up to give immediate height. There was a push at design stage to give real depth of planting, which would be carefully thought out and it had a realistic budget.

At about the same time, there was a reorganisation of gardening staff. This reduced some responsibilities for the horticulture department, increased the team numbers and stopped rotating staff between the animal and horticulture teams. For the first time in years, there was a permanent staffing level with more trained horticulturalists. During the first year after this change, the amount of work done was really phenomenal. It incorporated catching up on past years when just keeping ahead of the worst problems, such as weeds and hedges, was all that could be done. This was termed 'fire brigade gardening.' The difference was remarkable. Plus, it freed up time to look at the new possibilities that the developments would bring. There was an obvious attitude change. The plants were now as important as the animals. This was also reflected in the *esprit de corps*.

Initial plant lists for the masterplan had two parts. A short list of signature plants that were not intended to be used elsewhere within the zoo, only in that one area. A longer list would be of specific plants that would work well. The overriding consideration throughout was the appearance, selecting plants that looked right. All the better if they were hardy, tough, would take pruning if needed and were easily available in large sizes. Much time could have been spent sourcing really weird and wonderful plants and indeed some did come in, but the majority had to be easy and available in bulk if and when needed.

Each project had a different theme and it was very useful having these determined before too much plant sourcing started.

Asian Rainforest

Hide the fences, hide the buildings, make it very dense, a narrow visitor trail with no looking around corners. Basically, a bamboo forest, with some large-leafed, preferably evergreen trees and shrubs. These would break up the bamboo, which could have been overkill without an occasional change. Although the main planting had to be bamboo, thinking ahead to the African Rainforest was essential too. Therefore, when choosing for Asia, we only selected the smaller leaved genera, leaving mainly the larger leaved genera for Africa. Within any forest in nature, when a tree falls down it creates an opening, a sunlit glade, where smaller plants can thrive until the canopy fills in. Within the planting some more open areas were kept imitating this. These became the best spots for choice plants, even some from the Himalaya area, with viewing over them. Very interesting possibilities came to light here.

African Savanna

An open landscape with distant views and thin, scattered planting. No real hiding of fencelines was possible. What image comes to your mind when you think of the savanna? Dry grassland, with scattered trees? But in the rainy season there can be very colourful displays from a very wide range of plants within the grass and many of these are easily available common garden plants in good numbers. Trees and shrubs were chosen with small, often needle shaped leaves or resembling *Acacia*, the dominant tree in the African Savanna. This gave a very different looking range of plants. In places, some screening was needed.

Family Farm

The Family Farm was a farmyard with several animal areas very busy with visitors. It was a small project with little scope for planting and with huge risk of plant damage as some visitors love to feed the sheep and goats and cannot resist breaking pieces off. There was room for a few large trees and a long thin meadow of wildflowers with a traditional mixed species hedge to stop trespassing. Readily available native plants were used and for some of the trees, a cultivar thereof.

Gorilla Rainforest

This was certainly the biggest challenge in size, scope and plant selection. Our inspiration came from an area of the Congo called Mbeli-Bai, a marshy area in a forest clearing. A large area was earmarked for this project, with several mounds and a long water margin. Dense growth was wanted, both on the gorilla side of the habitat and the visitor side too, with viewing only in particular areas. Large-leaved trees were again needed, but different to the Asian rainforest. The large-leaved bamboos already listed were used and to make the area unique, some of the plants used were to have red or purple foliage. The visitor side was simple enough to list plants for, but the gorilla habitat needed a complex mix of edible, non-edible and non-palatable plants, and, it goes without saying but well worth repeating, nothing really poisonous. This would be a real test of the horticulture department's skill and knowledge. It took the best part of a year to assemble planting lists after much research, especially from colleagues in other zoos.

Sea Lion Cove

Northwest America's coastline was the influence for this area. It would be a most unusual planting as it mainly consisted of conifers, with an understorey of shrubs and loads of ferns. Dense screens were needed to surround the area, but they would take time to grow large enough. A bonus was that many of the plants would actually be native to the area we were taking our inspiration from.

Flamingo Lagoon

Working with water is always more interesting – and challenging. The recent experience working on the lake edges came in very useful here. This provided an opportunity to use dozens of one of the zoo's most noticeable plants – *Gunnera manicata* with their massive leaves on 2m stems. They had been rescued from the builders and potted a year before, which had been an interesting project in itself.

Asian Rainforest for Orangutans

An arboreal great ape needs high climbing to give the correct habitat structure. The engineering behind this called for very strong supports, hence concrete artificial tree supports were used. Planting took much experience from the gorilla habitat, but that caused further challenges on a very narrow island.

Asiatic Lions

After the initial build, this was modified. The inspiration came from the Indian Gir Forest which is very like a savanna, but less grass and scattered trees. The biggest issue was the constant mud caused by heavy animals regularly walking the same route. The first planting had some success. This experience was incredibly useful and could be developed further with some confidence that the planting would survive well.

In the space of ten years, from 2007 to 2017, the horticulture team of six, along with possibly one student and an extra pair of hands during projects, designed and planted the elephant, savanna, gorilla, family farm, sea lion and orangutan habitats. It was a quantum leap in the amount of planting, an incredible change in the appearance of the zoo and a radical change in visitor routes for viewing the animals. It was a great achievement by the small team of gardeners, using a lot of the little tricks learned along the way to not only make things easier but at times, to make things possible at all.

CHAPTER 5

Design and Thinking Ahead

Planting is our topic, but before any plants are even thought about, the bigger picture needs consideration. Most zoos will have a masterplan in place for future development in the ten years ahead, maybe more. Sometimes there will be a timeline of hopeful projects for each year, sometimes open-ended. Everything depends on many factors – the availability of the animals involved, the land needed, or, most likely, the necessary budget for the work. The masterplan is essential. It can save time, work (especially preventing or reducing the chance of having to do work twice) and money – if thought through carefully by the team involved. That team must be inclusive of all interested parties. You may need new animal areas, but you will need planting in and around them. You may also need visitor services, toilets, catering facilities or education buildings.

Before any of that, you will need to look at all services that may be needed, not just now, but in the future. These include water supplies (mains or rainwater, including storage), electricity, maybe gas/oil for heating (or if room for a central boiler, a combined heat and power unit, saving money and using timber as fuel), drainage (surface water and sewage) and pathways for visitors. Can a staff access road be put in separately if there is room and it will be out of sight?

Why is all this so necessary? Underground services are great, but very disruptive to put in, far better to have them largely in place before any large-scale building. It's simple really. It is always, always easier to put in main drains and ducting for other services – power, water, gas and communications – at the earlier stages and preferably plan it so that access for these is on a road surface rather than in a planted area. Drains may get blocked and cables may need replacing or increasing. It is far easier to work on a road surface than in amongst plants, with the inevitable damage to plants or requests to prune plants back. Remember, access is often needed for vehicles too. It is always easier on a road. Any traffic, even regular light foot traffic, especially in wet conditions, will adversely affect the soil causing root damage and/or soil compaction.

We should digress here for a minute. What do you call the 'area' an animal is kept in? A cage sounds awful. An enclosure sounds better, but that is really just a fancy word for a cage, isn't it? So, how about exhibit? It sounds more realistic, but could it perhaps mean the animal is on exhibit? Hmm…not so good… Eventually the term habitat became the default, which is technically correct as anywhere that an animal is kept is its habitat, though maybe far from the natural one. Maybe zoo habitat? A zoologist might consider the animals' habitat as where it lives, its home address. A zoo wants to make that address as interesting and stimulating as possible, firstly for the animal and secondly for the visitor.

As a horticulturalist, how can proposed short-term or long-term projects be looked at to determine where problems may occur that affect plants? How should these projects be looked at regarding planting design requests and working out budgets? Often, a simple checklist makes sure there is less chance of important things being forgotten.

Let us imagine a 20m square animal building for a small antelope, with 3m tall fencing to be installed around thin woodland to enclose 1,000m^2. The antelope may graze on any foliage they can reach, but not damage the bark of well-established mature trees. For simplicity, we will assume that this is already on the masterplan and there are drainage, power and gas (with appropriate connections) available for heating at a nearby visitor access road. There may be a concept drawing from the masterplan, which is best regarded as a basis for friendly discussion, but what is needed is a detailed drawing from the engineers. These drawings should show the building location, all nearby trees (which should already be on their own record system, tied in with the main zoo site drawings, with accurate positioning, showing condition, spread and height of tree and have regular reviews for safety), proposed paths, proposed fencelines, proposed drains, power, etc, plus all gradients. There should be an animal hard standing area for wet weather or winter conditions and for animal management purposes. Always regard these drawings as proposals – they are only lines on paper, often drawn up with no regard for any trees or tree roots, probably in a straight line to save costs and for ease of building – builders and engineers both love straight lines. You may make no friends arguing for changes, but you may save many trees. Plus, it is always better and cheaper to negotiate at the paper stage before the builders get started. Done carefully and quietly, you will gain respect from the team for thinking through the project before work starts and once you have their respect, the team will come back to you for advice and confirmation.

Let's consider the following.

Location

Where is the building? Out of sight behind a screen of existing trees? Are the trees evergreen or does this screen become see-through during winter? Maybe you need to plan in more screen planting, increase existing planting or plant a completely new screen? Could this be done even before building, to give the plants a head start? Will new access paths or roads be needed? Are trees in the way? Do you need to protect them from builders?

Existing Vegetation

There will be the risk of possible poisoning from plants, depending on the species of animal. You will need to check what is already there, including the trees and especially the surrounds, as plants may seed in or leaves from trees or maybe twigs/branches may blow in with windy weather. You should be aware of any previous use for the area, this may affect what soil depth is available and its condition. Is there any risk of any pollution of the soil, for instance from old oil storage tanks?

Soil

A subject in itself, soil is the basis for plant growth. It is to be treasured and nurtured if you have it, otherwise it must be bought in at some cost. You need to know whether you have clay soil that will be very difficult in wet conditions or sandy soil that may well need quantities of organic matter and lots of water too. Depending on what you want to plant, you will need to specify what depth of soil is needed and possibly drainage. If new tree planting is required, you'll need a big enough root area. In special situations, you may need special soil. Urban tree soil (a constructed mix of various sized round stones or sand with 20% air space filled with organic matter) in particular has many applications (see Chapter 14). You need to know your soil pH, how acidic or limey it is. This will affect what you can grow and might be important if asked to grow a particular range of plants. You may need to consider how to prevent soil compaction, especially for tree root protection. That will depend totally on the animal involved.

Almost always, you will not have enough soil on site. Topsoil is a valuable material. If you do have it, treasure it. Handle it carefully, keep it dry and keep the builders off it. If you have to buy topsoil in, make sure you do it or the builders use your supplier – a guaranteed way to ensure quality. Once on site, cover it with plastic sheets to keep it dry, make sure builders re-cover it if using it. Treble check that any areas ready for soil have had the subsoil de-compacted to allow drainage – machinery traffic or even foot traffic will cause compaction. Once it is buried and out of sight, no one else will worry about it, but the plants will not thrive. Once the topsoil is in place, it should be planted up immediately. If not, can it be kept dry, ready for planting and fenced off to stop builders walking, driving or storing material on it? Keep reminding the builders that the topsoil is the plants foundation and a poor foundation will give a poor plant – the same as the foundations for a wall or building.

Land-shaping

Depending on the animal, you may want mounds, slopes or a natural water course with plenty of rocks. If recycling the water, it is often easier to have pumps fitted outside the habitat for ease of maintenance. It will all depend on the animal and the habitat you are trying to create. It can be great fun making it all look natural, but always make sure the surface is safe for the animals.

Drainage

You may need to drain water away but watch carefully for unspoken plans to allow water to drain into planted areas. It may well be fine in summer, but will it cause waterlogging in winter? Soil will not absorb water if compacted, especially if dried after compaction, so along path edges where soil can be compacted by feet or wheels, there is little drainage and heavy showers can cause unforeseen problems. Can you drain to a nearby lake (if the water is clean enough to do so)? Can you drain to a simple soakaway or open drain or do you need to connect to a proper drainage system?

Sunlight

Where are you planting? Full sun, full shade or in between? This largely dictates what plants will do best.

Services

Knowing what services are already in place is essential before the building work starts. This is equally important for you if planting, especially when planting trees that may require heavy stakes being driven into the ground. Deciding with the builders where services can be laid is crucial. Keep asking for water points (as irrigation is much easier later) whether it will be a plumbed-in system on timers or a simple hose as needed.

Access

This is the most important aspect really, but one that is often not considered enough. Building work will need suitable access and entry for delivery and waste removal by large vehicle and staff will need a permanent way in, but also bear in mind animal delivery. Is there access for lorries or trailers? In the case of a larger animal, is there room for a crane to set up? Access to the animal habitat should include large enough gates for vehicle entry if needed, preferably hidden from visitor view. Imagine a large tree falling on a fence or a burst water main... Although in a perfect world, such services are always under pathways. You may need to renovate grass or add substrate. These tasks are always easier if large gates or sometimes an angled gate, is present to allow a lorry to reverse in more easily. A right-angled turn is never the best.

Plant Choice

So many aspects. So many questions.

1. Are you trying to give the impression of the animal's natural habitat? Which plants will do that? Could you use some plants from the animal's natural range?

2. How much damage to the plants will these animals cause, if any? This depends on not just the animal species, but often the individual animal; some are docile, others the reverse. Damage may depend on stock density – two may do no harm, ten probably would.

3. Are you being asked for an instant scheme using big plants or can you get away with smaller, cheaper, easier-to-plant sizes which will need far less checking or watering? Maybe a mix of the two would work?

4. Planting time. This is always an issue. Builders will not want you in the way, nor will you want the builders to have the chance of damaging new plants. Planting will almost always be the last thing done, often with very tight timelines. Note that this is not the best way to do it.

5. Are the animals expected to eat the plants? With many plants and not too many animals, the animals eating the plants is not ideal. Preventing this can be achieved successfully with distractions such as extra scatter feeding or extra browse put in to allow the new plants to establish, but problems often arise.

6. Do you need to plant to provide a screen for the animals to hide from sun, visitors or each other?

7. Can you plant for enrichment, inside or outside the habitat, for attracting insects or an occasional nibble of the plants themselves?

8. During all plant conversations, you need to think possible toxic, hazardous or irritant issues. This can be a massive task, especially for herbivorous animals. What is poisonous and what is not? Is it unpalatable so not touched, although poisonous? If so, is it safe to risk it? This can vary between animals in any one species; a newly introduced animal may find unfamiliar plants and eat them on the first day not knowing it shouldn't or are they intelligent enough to be careful?

Viewing

Most plans will fix a viewing area for the visitors, dictating the view, ensuring features are hidden, such as the housing, fences, etc. The planting needs to allow for this, with viewing through plants, maybe raising the crown of shrubs or thinning out some branches, very careful hand pruning looking at every cut or using low-growing plants that can be seen over. 'Over' means making the habitat visible to anyone sitting in a buggy or a wheelchair or young enough not to see over half-metre high vegetation – just because you can see over, does not mean everyone can! The only way is to kneel down and see what you can see or go around in a wheelchair and test the system for yourself.

Furniture

This is a catch-all phrase. To make it look more natural, you may need to include large rocks (which need to be placed early with heavy machinery), logs, roots of trees – all in careful agreement with any mounds or water features planned in. The animal team really should decide where things go and what is needed. They need to think of feeding areas, maybe hiding food daily, scatter feeds to encourage searching for food, walking areas and probable paths – many animals will establish a regular walk or path in a fixed area. They need to be aware of issues such as animals hiding from view, climbing high on logs when too near fences or digging. There may be carefully made artificial rocks, termite mounds or even trees with a system of feeding to encourage natural behaviour.

Substrates

This depends on many aspects. You may be aiming to recreate a desert habitat using sand or grit. You may want the animals to be able to dig for insects which you can bury first. Bark or wood chips can give a soft effect. Check what you are using though. There are many different types on the market that may be less suitable for animals. Wood chips may be fine, but from what tree? You can source specific tree bark in some countries, with very specific acidity and bacterial action. Is the tree toxic, such as yew *Taxus baccata* or elder *Sambucus nigra*? Does it have thorns like hawthorn *Crataegus* spp? Is the sap sticky like some conifers, such as pines *Pinus* spp or western red cedar *Thuja plicata*? Most hardwoods are fine but the sap from horse chestnut *Aesculus hippocastanum* can wash out of chipped timber with rain and it floats on water forming an oily looking puddle. That is when you remember it is in the soapberry family which often have saponins. These are probably harmless but scary looking on water, like an oil spill or diesel leak.

Inside Habitats

A whole new area of potential problems – sorry – challenges. Very often these are too small for effective planting, but if there is room, there is often plenty of scope for a good range of plants. The overriding consideration is light. Perhaps additional artificial light can be used, but it can be hard to get it right or near enough to the plants. What temperature range will be set and most importantly, what humidity level? Can this be boosted if needs be? The heating system can adversely affect the humidity too. All these systems have to be animal-proof too – no mist nozzles to be played with, no pipes to be pulled at or nibbled. A large greenhouse can be a very nice habitat and there are some stunning examples around Europe, but success will depend on the animal species chosen. Some small primates may be alright, others less so. Many birds will do well, but parrots, macaws and parakeets can chew everything at any height or angle. Be extremely cautious.

Budget

The last point, but the most significant perhaps, is budget. Often a costing for planting is expected after a simple concept drawing is passed around. This is not ideal but can be useful. Get a rough idea of planting area, estimate the required soil type and quantity and then add an extra 50% just to be sure and give your estimated figures as a baseline. Then if any variations come along, you can say with hand on heart, 'Well, this has changed, so much more plants will be needed...' For planting costs, look at planting density and the size of plant. Nine herbaceous plants in 1L pots per square metre will cost very little, but three bamboos in 12L pots per square metre will cost an awful lot more. Each tree will obviously vary in cost depending on species (some are easier and quicker to grow) and size.

Builders

Builders' access will hopefully follow the intended keeper access path. Other access paths may need tree root protection, spreading the load of vehicles or preventing damage from tracked machines. Consider fencing off all non-builder areas to prevent the accidental driving of vehicles, tipping of material, parking of vehicles or storage of material. Strictly control any waste disposal, especially the cleaning of concrete mixing machines and look for diesel spillages. There are very thorough guidelines on limiting damage to soil and planted areas available. You just need to look at the programme of work with a horticultural eye. The plants cannot shout for help. Potential issues need to be spotted – hopefully before they become a problem and that keeps everyone happy. Any pathway construction to protect roots has to be done before the builders start work. A lot of damage can be caused in the first few days with heavy machinery and poorly judged access.

Trees

By carefully choosing the actual position of the house, you can determine whether there are trees in the way. Are some of lesser value? Can damaging the very best trees be avoided by sacrificing diseased, damaged, older or less valuable specimens? For instance in the outside habitat, can some less important specimens be left as sacrificial trees, as scratching posts? Allow for builder's access to be greater than the building site. If felling trees, can some be left in situ to make the habitat more interesting for the animals or visitors? Could some trees be pulled over, leaving the root plate vertical as a natural windbreak? Can any felled timber be left nearby, outside of the animal habitat for natural decay, insects and fungi, and to provide a natural habitat for native fauna too? Perhaps safely leaving some stumps still vertical with a limit on their height, for insects and insect-eating birds might be a possibility? Are any branches or rotten timber useful in other areas of the zoo for habitat enrichment or natural perching? Any tree work will generate wood chips from shredding the unwanted branches. This material is perfect for mulching newly planted trees and shrubs. Can it be kept for later use? If not, it is good to just let it rot in thin layers in nearby woodland. Be very careful not to allow any heaps to form around the base of plants as it will heat up with bacterial action initially, especially if there are many leaves are in it. This can damage the bark and cambium and it can kill mature trees.

Access

All routes of access must be checked – putting in path or road foundations will damage many roots. Can the path be carefully but simply woven between the trees? Access may be needed for trailers; this can be awkward on corners, particularly reversing in or out. A road or even a simple pedestrian path, is often a standard format – dig out the soil, place stone foundation, lay tarmac. There are other ways to lay any path or road. A stabilised foundation using a cellular grid, either rigid or an expanding cell format, can be laid directly on the soil surface. Be aware this goes against most builders' experience. You may need to be there to make sure nothing is removed first or a digger may go in and strip some soil along with any roots that are there. Befriending the builders and the drivers, visiting daily if you can, letting them know to call you before any such work is done may prevent damage. For a stronger foundation, use a deeper cellular mat or simply lay two layers – simple and very doable.

Hard Standing

This is an area with a solid or sand surface for multiple uses in winter, in wet weather or for managing the animals. The animal team will want a surface that does not become muddy in rain. Concrete or tarmac may be too hard for the animal's feet. A stone surface may be specified to allow drainage while also allowing some wear on the feet as a natural process. Due to their nature, these surfaces are very hard to clean by animal staff as small particles from manure or autumn leaf fall will accumulate surprisingly quickly to form a thin impermeable layer on the surface, causing water to lodge. All of this spells trouble for any tree roots – with waterlogging on the surface, there is less air getting into the soil, which is essential. In time, tree roots will fail little by little, the tree suffers and eventually, the tree may completely die.

Soil Compaction

This is probably the most common reason for tree problems on any soil surface after building and with animals. To prevent mud or to enable cleaning, there are many preformed systems available, but nothing is perfect for the trees or for the animals, let alone something that suits both well. There is one system we tried a few years ago that was very much outside the box and has worked extremely well for several years now in both the bongo *Tragelaphus eurycerus isaaci* and okapi *Okapia johnstoni* habitats in Dublin Zoo. Both have a hard standing area with no vegetation except mature trees and a habitat with trees and grass too.

Having lost many mature trees to compaction, the topsoil was rotavated when dry enough. This often revealed a surprisingly dry soil underneath a very thin and muddy compacted surface. It is amazing how a thin layer of compacted soil is so waterproof. On top of this rotavated soil, a strong plastic mesh normally used to reinforce grass surfaces for car-parking was laid. Usually this is pegged down, but with worries about the pegs coming up and potentially damaging feet, cable ties were used instead to connect the mesh together and they were carefully trimmed off. Approximately 50-75mm of stone – urban tree soil – was laid on top of this, but a very deliberate mix was used. The stone size was previously checked and agreed on with the animal team. Only round stone was used, so no sharp edges.

Urban Tree Soil

It is worth briefly describing the principles of urban tree soil. This is normally used as topsoil replacement, but was used here as a thin layer or a mulch on top of existing topsoil.

Imagine a bucket full of round stones. They cannot be compacted firmer than they are in the bucket. Now mix different sized round stones. Again, they cannot be compacted firmer. You need approximately 20% air space for urban tree soil, which in the final mix is filled with organic material. If the bucket holds five litres, just add water until it overflows, but measure how much goes in – this equals your air space. With 20mm stone that would need two litres of water and with only 6mm or only 10mm stone it still needs two litres or 40% – giving too much space. But if you mix stone sizes together, the air space reduces. With a 50/50 mix of 6mm and 10mm stone, that air space becomes the magic 20%.

Into this 50/50 mix of 6mm and 10mm stone, add 20% of partially rotted wood chips from tree surgeons. The stone allows drainage. The mesh spreads the load to prevent compaction. The tree chips hold a little moisture and help grass seed to germinate and establish. This worked incredibly well in the okapi and fairly well in the bongo habitat. Germination was less in the bongo habitat which was likely caused by the hoof movement while the animals walk. The same urban tree soil mix was used in the Asian lions and Sumatran tigers' habitats. It was laid on top of normal topsoil. This worked very well, as it drastically reduced the muddy conditions. Wherever the animals walk, a path will develop. There is bound to be some compaction of the stone into the soil and mud will appear. A simple solution is to add a little more stone on top. Whatever you do, do not add wood chip on top, as this will rot in time and cause more mud, not less.

CHAPTER 6

Cooperation

Learning From Peer Groups

Before new designs and individual project details are discussed, let us look briefly at how zoos around the world work together. As with many things, cooperation benefits all. Various areas around the world have local zoo associations. The British and Irish Association of Zoos and Aquaria BIAZA is Dublin Zoo's local group and they have certain criteria for membership. They have very strict rules and regulations as to how a zoo is operated, with annual inspections. This dictates some elements of the animal areas, animal management systems, fence heights and construction, the size of internal areas, visitor stand-off fence heights, etc. There are meticulous records kept of animal breeding. As in horse and cattle breeding, these are called stud books and are international. For each endangered animal species, there is normally one zoo that keeps these records. This spreads the workload and helps to keep the genetic basis of the animals held within the group as diverse as possible, a most important aspect of species survival.

The annual inspections do not have much input into or effect on the horticulture aspect, but it would be usual to look at trees from a safety aspect, work which would be normally done anyway. Having an up-to-date spreadsheet including tree work carried out, tree number, date of work, reason for work, is simple enough to set up and maintain. It is very handy to have for inspections if needed. After looking around, inspectors have asked about the 'invasive' *Gunnera* growing around the lakes which of course, was not the invasive *Gunnera tinctoria* but only the well-mannered *G. manicata*. A full list of all *Gunnera* species grown was given with descriptions – *G. perpensa*, *magellanica*, *manicata* and *kilipiana* and no more questions were asked. That proved the value of an up-to-date spreadsheet of all plant accessions, planting area, date, source, etc.

Within Europe there is a larger group, the European Association of Zoos and Aquaria EAZA. They have guidelines and more recently, they make inspections too. Above them all, there is a World Association of Zoos and Aquaria WAZA. They tend to look at overall direction rather than specific points and give excellent guidelines. Within both BIAZA and EAZA, there are specialist groups that look at particular animals or maybe groups of animals, primates, reptiles, etc. This has been the practice for many years and has helped in many animal management areas, sharing information for the common good of the animals.

In the mid-1990s within BIAZA, a horticulture conference was organised by the Curator of Horticulture at Edinburgh Zoo. This was the very first time a conference had been planned purely about horticulture in zoos. It was an excellent initiative and incredibly useful. There was much interest from the twenty-five attendees and plans were made to make it an annual event. This happened and proved so very useful. EAZA saw what BIAZA were doing and encouraged a European meeting. For many years, there have been annual conferences all across Europe, from Barcelona to Budapest, Rome to Randers (Denmark), Zurich to Dublin and many more. These events have proved so useful and greatly beneficial. They have been a great opportunity to see what other people are doing and especially to see new plants or planting ideas, which is always good.

Within BIAZA, several Curators of Horticulture formed a Plant Working Group to fit into the BIAZA organisation structure. In the first year of working together, the group were asked to look at writing a chapter for the Animal Management Course for animal keepers. This was a partnership between BIAZA and Sparsholt College and is an excellent grounding for all animal keepers in Ireland and the UK, giving them a great start in all aspects of keeping animals. Dublin Zoo Horticulture Department volunteered (when will it learn...) to try and write the chapter on zoo horticulture, with other zoo colleagues peer reviewing. It was too good an opportunity to miss!

All zoos had seen plants killed, treated badly, pruned terribly or over/under-watered. The course chapter became a basic 'how to' gardening book specifically for keepers. They knew that plants were good for their animals for many reasons – shelter, screening, climbing, attracting insects in, occasional nibbling, etc., but had no idea of the best way to grow plants, how to prune, even which would do best and, most importantly, which plants to avoid. Results were often very poor or looked awful, especially to a gardener. There was also a lack of knowledge about weeds. Many zoo animals share their habitat with a great collection of plants that visitors recognise as weeds, especially tough nettles *Urtica*, thistles *Cirsium* or dock *Rumex*, any of which give a neglected appearance as if the area is not cared for – not the right impression at all. Simple selective weed cutting and removal (editing) and the stopping of seeding all contribute to an incredible improvement in the visual appearance of the habitat. It also reduces the spread of weeds. Surviving weeds could indicate the degree of animal damage to plants, as only the really tough survive in many animal areas, particularly if too densely stocked.

With the new title, *Diploma in Management of Zoo and Aquarium Animals*, the course became bigger and better with regular revisions. Having a whole chapter specifically on growing plants in zoos has been of immense value over the years, as each new keeper has to complete a chapter focusing on what plants to use. Just think of that – each animal keeper has to think about plants. What a change. It is such a basic element of animal life, but it has been a real eye-opener for many animal keepers. They have to consider not only what they want the plants to do for the animals – screening, cover, shade, insect attracting, maybe a little browsing, but also how to plant and care for them – watering, pruning and everything. Habitat is the whole package. Plants and the ecosystem they build so often make the habitat. The animals do not live in a cage or enclosure but in a habitat, so any activity for maintaining the plants is called habitat management. It is all gardening, but first and foremost is the link to the animals as to why it is being done.

Pooling the experiences of many zoos led to a spreadsheet of plants in various animal areas in several zoos. This spreadsheet rapidly became unwieldy, so a web-based database was developed instead www.zooplants.net. Initially, this was largely implemented by the Royal Zoological Society of London, London Zoo, with input from other zoos. It is now the most useful collection of data about plants within a zoo environment, enabling searches by animal or plant. Therein you will find practical experiences from many zoos, with details about successful plant uses, risks of poisoning, and much information about nutrition based on various browse plants at different times of the year. An absolutely priceless resource, it is often the first stop when looking for ideas. There is always a caveat that must be remembered when dealing with animals and plants – just because it works in one zoo, it may not work in another. There are many reasons why. These reasons include: the group dynamic of the animal may be different, there may be more animals and less space, the wrong substrate, a more destructive male that likes to show off by damaging plants or, more likely, a new adult addition that is not used to plants. There are several risks to such animal movements. We might worry that the plants will get damaged but, far worse, is if the animals are not familiar with plants, they may eat something they really should not. If the plant does not taste nice, they will probably not eat it again, but if it is poisonous, they may die – that's a very different issue. Hence, great care is taken about what is planted – whether edible, unpalatable, poisonous or deadly poisonous. The difference between the last two is awkward as many poisonous plants taste bitter and will not be eaten much. The really dangerous ones may taste ok and not have a bitter warning.

CHAPTER 7

Asian Elephants
Kaziranga Forest Trail

The Asian elephants *Elephas maximus* project, the Kaziranga Forest Trail, was the first large landscaping project within the new masterplan of themed habitats in the zoo. There was a lot of discussion about how to do it and its design, with no really good result. Eventually, the task was given to Grant Jones and his team from Jones and Jones, Seattle. They are a company with much international experience designing zoos and other public areas. This was a real game-changer. They were not afraid to suggest removing this, knocking down that, changing this road, raising here, and lowering there. The design was the driver, especially to make it interesting for the animals and the visitors. The cost was often secondary, though sometimes design did have to be modified to match the budget.

The existing elephant area had been increased a few years before when the giraffes and zebras had moved up to the new African Plains. For this entirely new elephant area, the beavers and otters had to go too. And the bat and parrot house. And the old octagonal aviary. And the old restaurant extension to the Haughton House (another future project on a medium finger). It was the first time so many small areas had been incorporated into one and had the added benefit of actually reducing the number of visitor pathways. This is a problem often seen in older zoos where there are so many visitor paths, you lose animal space and have a very inefficient viewing route. This can make it easy to miss some areas and be confusing to walk around.

The design, a meandering trail through dense vegetation following a stream, had one awkward problem – the expected main entry for the visitors was at the Haughton House end of the trail. The stream would flow from there, starting at a waterfall, and accompany visitors as they walked along. Just one problem – it would be flowing uphill. Ahhh, yes… That's ok though. We can make the stream deeper at the far end. Then the plans were looked at again, and the realisation that the far end of the stream would be 4m deep dawned – a bit too deep to accompany anyone. Back to the horticulture departments' suggestion to make it two streams. By raising the level of the second half of the stream, halfway along in the middle where the wee bridge crosses it, no one would see it. Two of everything were needed – two sumps, two circulating pumps, two dirt traps. It also meant there would be two areas to keep clean, but the stream would look good. Plus, no one would notice the height difference in the middle.

The real complications arose, as usual, because we were dealing with live animals. The new elephants were coming from Rotterdam Zoo, where the herd was too large. The plan was to split the herd. Elephants are matriarchal, so we had to have mothers and daughters. To make it better for Dublin Zoo, it was planned to make sure they were all pregnant on arrival. This would save us the trouble and cost of getting a bull elephant for a few years. But it is always a risk moving animals, especially when pregnant. The plan was not too early and not too late in the pregnancy, so we had an eighteen- to twenty-two-month gestation to plan around.

Plant selection was left entirely up to the horticulture department – an incredibly free hand. A budget was put together with totals for trees, bamboos, and shrubs. There was no real concern about being accurate, to be honest. If an invoice came in for *Paulownia*, *Phyllostachys*, or *Petasites*, no one in accounts was expected to argue whether they were trees or bamboo. The budget was determined by calculating a rough cost per metre. Herbaceous plants were the cheapest, then shrubs, then trees, and bamboos were the most expensive. A little indulgence was alright, once still in budget. Irish-based nurseries had been used where possible for years, and with this much larger project, that practice continued. It paid off later for several reasons.

What plants to use? Our initial plans, as part of the masterplan mentioned earlier, steered us towards bamboos and large-leafed evergreens. Now we could look at specific choices. A range of bamboos would give great variation in height, growth habit, leaf size and shape and culm (stem) colour. Every bamboo has a different shade of green, with different requirements along the trail, a wide range could be used.

The plans were in place; the plant list was sorted, the plants were ordered for delivery in late winter. There was a problem with the elephants. Delays were talked about. It was decided to put the whole project back twelve months. All of this with thousands of plants about to be delivered. Fortunately, bamboo was two-thirds of the planting and could be held in the production nursery. The rest of the plants were delivered and potted on – taking over a quarter of the staff car park as there was no room in the then very small nursery. This worked incredibly well, giving larger plants for planting the following year.

Detailed plans were drawn up and shared. A stream flowing through a bamboo forest, with a waterfall at the start, and the visitor trail made of concrete fashioned to look like a muddy path with leaf prints and elephant footprints, which would be great for kids to see and wonder at. The waterfall and many solid barriers, were made of concrete too, but very artfully. The forewoman on that part of the job was actually an incredibly talented artist, working with an American company that specialised in such work. For each concrete wall that was needed, the builders poured the foundation with lots of steel rods protruding to give a secure fix. The artists then moved in and used the reinforcing steel rods to make the shape. Steel mesh was wired to the framework, and hessian was put in place behind that again. Visitor-side areas were made more natural and interesting by wiring in large stones or tree roots. Trailer loads of roots were obtained with great difficulty from an area of bog where birch *Betula pendula* had been extracted.

Concrete was fired through a hose at the framework – literally like water – but using a much heavier hose with a specialised pump. As the concrete builds up, the hose is aimed at the next area, allowing only so much at a time and allowing each area to dry. Layer upon layer was added, depending on the need. If it was to be used for soil retention, more thickness was required; this also depended on the height. Many of the walls were circular. It was similar to making a very large pot in places. Much of this concrete would be visible, and the intention was to make it resemble a mud wall, where the stream had cut through. Colour was added to the concrete mix to help this imagery. Later on, any visible wires from the stones or roots were removed, and any concrete on the large stones was carefully taken off. Then the artist side of the team added in detail in the form of colour flecks here and there. In natural soil layers, the top 250mm or so is the most biologically active. It has the most rotted and semi-rotted organic material in it, which means it is often a slightly different colour with more brown or black than the subsoil below. The artists painted the concrete to reflect that – a rich organic colour for the top 250mm. Remember, the wee stream bed you are meant to be walking down has eroded the soil. Hence the visible layers.

Entrance to trail, fifteen months after planting, two summers growth. Compare to the two pictures above, with the cedar and *Sequoia* for reference. The waterfall is completely hidden until the corner is turned.

Large-Leafed Trees

Ten cheap and tough London plane *Platanus x hispanica* at 5m tall were used to give immediate height. This hybrid plane is very tolerant too, hence planted so often as a street tree. Some of the planting areas were smaller than we would ideally prefer for a tree with pathways nearby. This tree would survive, though maybe not thrive. A terrible thought was that we might need to cut some down if they all grew well. It is awful to say, but some were planted as sacrificial; in a few years, they may need to be removed in favour of a more unusual tree.

The Japanese walnut *Juglans ailanthifolia*, with leaves more like the tree of heaven *Ailanthus*, but with the longest walnut leaf of all (600mm to even 900mm at times), were wonderful.

Chinese necklace poplar *Populus lasiocarpa*, is the largest leaf poplar, with leaves of up to 250mm long, enhanced by the lovely red stalk and midrib to the leaf and with masses of downy seeds flying around early in each summer. Magnificent tree.

Foxglove tree *Paulownia tomentosa*, so-called as the flowers resemble foxgloves, and for years it was in the same family, Scrophulariaceae. Though now it stands alone in Paulowniaceae. The flowers are borne in May but are formed the previous summer and exposed all winter; bad frost can destroy them all. Ooh, but when they do flower... Imagine a purple foxglove flower, 50mm long, with an open bell-shaped mouth and a wonderful scent. You may need to find a fallen flower, pick it up to enjoy it. Glorious. The leaves are normally up to 300mm across, but if the plant is coppiced (cut down each winter), the new growth will give leaves easily double that, and the shoots will easily reach 4m in a year. This was tried, but the surrounding vegetation, especially bamboo, was too dense, and the new *Paulownia* stems grew weak and twisty. Some were left to become trees, even if they were leaning badly as it looked more natural. They are great when they flower, and when in leaf, they look so tropical. If another opportunity arose, it would be best to plant a dozen in a wide clearing with plenty of light and feed the brutes.

Paulownia tomentosa flowering on a low branch for a change, normally seen against the sky. Fruiting *Fatsia japonica* behind, a great food source for birds in spring.

Shrubs

The preferred option would be with large leaves, evergreen if possible, or maybe with an interesting highlight. Luckily, there are many to choose from. Many of these were planted among the bamboo, a lot were lost – simply outgrown as expected – but where they have done well, it is great to see the highlight of a different leaf colour, a red berry or even a flower or three, amongst the bamboo itself.

Heavenly bamboo *Nandina domestica* is not bamboo; in fact, it is related to *Berberis*. The Chinese name for it is 'nanten,' hence *Nandina*. The doubly pinnate leaves are bright red when young and dark red in autumn and it gave a great contrast within the green-leaf real bamboo. Seed heads of bright red berries stand out equally well.

Wrinkled or leatherleaf viburnum *Viburnum rhytidophyllum*, so named because the leaves are deeply veined, is from China. It is a gem, but like many shrubs, looks better when growing vigorously. The best specimen has now been crown lifted, so you walk under it a little, while giving more room for herbaceous planting underneath. The large leaves are really ornate, even architectural, while the top surface is a deep lustrous green, the underside has a furry light brown covering. Flower buds formed in late summer are held all winter, giving an interesting appearance.

Harlequin glorybower *Clerodendrum trichotomum*, again from China, is a most unusual large shrub and why wouldn't it be? It is either in the Lamiaceae, or the closely related Verbenaceae family, depending on which authority you read. In northern European gardens, these are normally herbaceous plants but once you get into more tropical climates, quite a few herbaceous families have woody members too. The flowers are interesting – white, star-shaped and slightly fragrant, but the season of interest is much extended with the red calyces remaining in place. Further colour comes from the berries – white at first, then shiny light blue and eventually, turning a very dark blue. These are all good colours in autumn. The leaves are strongly scented of peanuts, though unpleasant according to some people.

For large evergreen leaves, it is hard to beat the southern magnolia or bull bay *Magnolia grandiflora* from North America. Usually seen as a wall shrub, here it is planted in the open, though it is quite sheltered by other plants. Similar is Delavayi's magnolia *Magnolia delavayi* from southern China. Large white flowers of up to 20-25cm across appear on both during summer and are both spectacular and a pleasant surprise among the dark green leaves.

The chestnut leaf holly *Ilex x koehneana* 'Chestnut Leaf' (a hybrid between the European holly *I. aquifolium* and the Japanese holly *I. latifolia*) is, as you would expect, a large-leaf holly with a leaf up to 15cm similar to sweet chestnut

Castanea, which grows fairly slowly up to 8m. It is another good solid reliable evergreen between the bamboos.

From the outset, the planting list included the Japanese aralia *Fatsia japonica*. This is a wonderful shrub with great palmate leaves of up to 300mm across. It is very reliable, tolerant of shade and dry soil, a real stalwart, and does not mind heavy pruning if needs be. New shoots often come from the base. It is a member of the ivy family, Araliaceae, which is evident when you see the white flowers and the black seed heads just like ivy, but on steroids. Great for pollinating insects, as are many other Araliaceae.

More colour came in spring with *Photinia x fraseri* 'Red Robin.' This is a common plant, but still welcome. The new red leaves really show up well against the bamboo.

Chilean myrtle *Luma apiculata* slipped in quietly. This is a small evergreen with scented leaves, small white flowers followed by dark purple berries, but the irresistible feature is the most beautiful orangey-brown bark. Planted very near a path and gently crown lifted, this will eventually – hopefully – be a real feature amongst the green background, with room underneath for herbaceous plants.

Salix magnifica has the most un-willow-like leaf – when first collected it was thought to be a *Magnolia*. The catkins can be up to 250mm long. It is a lovely small tree.

Vigorous bamboo rhizomes crept into the sand of the elephant habitat, to be pulled out by the elephants. The young elephants found they could access some of the plant culms if they laid down. The parents waited for a few leaves to be pushed their way. This was the visitors view for a few days until a different fence system reduced the behaviour. Good enrichment behaviour for the animals though...

Herbaceous Plants

Imagine a wild bamboo forest – tall, dense vegetation, which would have little room for smaller herbaceous planting, except perhaps where trees had fallen or elephants had been browsing, creating 'openings' in the canopy. Some areas were left bare of bamboo along the visitor side of the stream. With a couple of large spaces available, these became our hotspots – imitation openings in the forest for a little colour and a more varied plant selection.

When flowering, giant Himalayan lily *Cardiocrinum giganteum* is a 2-3m statement of 'I'm here!' that no one can ignore. Many had been grown from seed five years previously, but there had been nowhere good to plant them. It takes up to seven years to flower from seed, and here was the perfect spot. Fifteen went in. After the really bad winter of 2009/10 when the horticulture department was very depressed because of so many dead plants, the *Cardiocrinum* was seen starting to flower. It cheered everyone up no end. The large white flowers of 250mm or so, with a dozen flowers per head, shine amongst the green of the bamboo. The bulb dies after flowering, but daughters keep up the annual display, though it has never been as good as that first year. Replanting will be needed at some stage.

Young growth in spring of *Cardiocrinum giganteum*, with the obvious flower stems promising a great show. Such verdant growth is always welcome after a long winter and it's a good healthy glossy green too.

The flower stems reach 3m – you really look up to these flowers, a marvellous show, even if they did screen the elephants a little. If people complained, the horticulture team promised to cut them down – not!

Umbrella plant *Darmera peltata* (on most wet days, I think I would prefer something bigger such as *Gunnera manicata*) delights with a simple pale pink 1m tall typical *Saxifrage* flower in early spring from bare soil. But oh boy, the leaves take over in summer, forming a dense cover of, well, umbrella shapes. It is a good green too. The 25mm thick rhizomes are surface travellers, looping in and out of each other, making a very dense mat. These are excellent for water margins to keep soil in place, and the plant loves water too. It does spread. After a few years, it was removed to make room for choicer plants and to make visibility better at low levels. Digging it up was rather interesting and heavy going – great use of a mattock. Those layered rhizomes meant it needed to be dug up small bits at a time, many of which were teased apart and potted on for use elsewhere. From an initial ten plants, after nine years, we had 150 x 12lt pots and lots left over. The wee stream along the trail actually sprouted a few seedlings of *Darmera* – literally growing on the rough-surfaced concrete used to make the stream bed. They were happy but small with a mat of fine roots holding onto any organic debris and the daily water flow.

The biggest herbaceous plant was *Gunnera manicata*. This is just one year after planting, with *Rheum palmatum var. tanguticum* just behind. The ducks were happy too. Compare to picture two taken during building.

The area near the waterfall was one of the best open spots for planting. *Darmera peltata* grew extremely well. White *Lilium regale* gave a good show. Other lilies, *Primula* or *Hosta* had their moments too.

Himalayan ginger *Cautleya spicata* 'Robusta' is a wonderful small member of the ginger family, though not planted often enough. Originally from the Himalayas through to China, it likes a little warmth to start growing. Just don't expect much to appear above ground until May when the good green leaves look grand and are soon the background to unusual flower spikes, red bracts with yellow sepals in abundance. The seeds appear as white pods, but it is actually a white aril attached to each small black seed. Seedlings grow readily.

Cautleya spicata 'Robusta' in full flower.

The ginger family also gives us purple roscoea *Roscoea purpurea*. This is another charmer, but with a shorter period of interest and a need for a little warmth to really get going. As with *Cautleya*, this is its nature. Coming from Nepal, it is used to cold winters. *Roscoea purpurea* has good purple flowers that can be likened to an orchid until you look closely. It gives a very pleasant, although brief, show in season. Interestingly, in Nepal, it is apparently pollinated by only one species of long-tongued fly, yet there is still pollination as seed is set here too.

The ginger family includes ginger lilies (though not really lilies) *Hedychium* from India. It is a really great tropical-looking gem, with large leaves and splendid flowers at the top of strong shoots. Unfortunately, they also need warmth to get going, and they flower late in the summer/early autumn for most species and cultivars. After a cold spring and slow summer, you can have the flowers very late. Some of the planting pockets were in part shade, so cooler. The best one seemed to be *Hedychium densiflorum* 'Assam Orange.' Originally introduced by Kingdom Ward in the 1930s, this is a good strong orange colour, and more importantly, it flowers a bit earlier. A reliable show, even if left in the ground. It probably should be mulched to keep warm in winter.

Real ginger went in too, Japanese ginger *Zingiber mioga*, another great foliage plant, but don't look for the flowers on top as they are borne at soil level. You may need to know it is in flower to actually spot them. The flower looks very similar to

Roscoea, with the same bright red inside seed pod colours of *Hedychium*. Though not a great display plant, it is intriguing, nonetheless.

No shaded woodland walk in Asia would be without some *Primula*, so various Candelabra primroses went in – *P. x bullesiana*, *P. beesiana*, *P. bulleyana*. They all looked great and intermingled. They were very colourful in season, whatever their name. As with the *Darmera* above, some have seeded onto the concrete stream bed, forming a mass of roots around any small debris that collects.

Lilies were an obvious choice when looking for bright highlights of colour. Many varied species, hybrids, heights, and colours are available, but they are very colourful, almost too much so, and perhaps not suitable here. The term 'gardenesque' comes to mind again, making it appear more like a garden than a natural habitat. Tiger lily *Lilium lancifolium* fitted the bill perfectly; a strong orange colour to catch the eye, but it is a simple form, not too large, vigorous enough to do well. Originally from the Himalayas through to China, so a real Asiatic plant too. The final tick in the box is the incredible ease of propagating them. Every year a small bulbil grows in the axil of every leaf. In nature, these would fall off and gently spread the plant around. It is very easy to collect the biggest and best bulbils from the strongest plants, pot three or four to a 9cm liner pot, grow on for one year, then plant where wanted. If only some grow, it would still be enough.

Lilium auratum var. *platyphyllum* is a real eye-catching, large-flowered lily.

Lilium lancifolium has bright orange flowers guaranteed to catch the attention of visitors.

Lilium lancifolium is popular and easy to grow and propagate.

Lilium nepalense is a real stunner, sumptuous colouring.

Roscoea purpurea has almost orchid-like flowers.

One rarely seen plant was used, the giant Japanese butterbur *Petasites japonicus* var. *giganteus*. This is potentially a real monster. The roots travel, and the rhizomes (even the flower buds) are able to push up through tarmac. Isn't nature wonderful… On the trail, it was contained in a narrow bed surrounded by concrete against the elephant pools. Considering how little soil it had, it grew well – too well. Annual pruning in midsummer was called for, otherwise anyone under 1m tall could not see over it. The problem was solved by the bull elephant. He learned to walk in the pool on his hind legs. Around the pool edge he went, his front legs on the concrete edge itself, leaving his trunk free to pull the leaves off. With such regular browsing, the height was reduced, meaning less work for the horticulture team. However, the area is not as verdant looking as it once was.

Large-leaved *Petasites japonicus* var. *giganteus* grew extremely well in front of the elephant bull pool and needed reducing in height each summer to allow views into the pool – until the bull elephant found a way to harvest the leaves.

Return to the Budget

So, what did the landscape cost? Fortunately, the builders' costs of excavating the areas for topsoil and placing the topsoil were part of general costs. The horticultural budget was only about soil, plants, and irrigation. Labour costs were the department staff, plus one extra student. Because of its slow growth to saleable size, bamboo is the most expensive at about €60,000, but well worth it. Trees, shrubs, and herbaceous plants would have been less than half that. The shock to most people outside of the horticulture team was how much topsoil was needed. The bill for the topsoil came in around €50,000. Yet again, the argument of topsoil being the foundation of the growth came up. At the end of the day, money spent was not that far off the original budget for planting – once you included the add-ons not requested until halfway through the build.

Orange flowers show up well in a green jungly tangle. *Lilium lancifolium*, *Hedychium densiflorum* 'Assam Orange', *Aristolochia tomentosum* scrambling over the bamboo and *Darmera peltata*, all against a green bamboo background in a sunny glade.

Acacia dealbata brings a great show of colour in late winter. This is eight years of regrowth from the roots after severe frost damage, when the entire tree was killed to ground level. With better soil and more water, growth is always very quick in Ireland compared to their natural range in Australia.

Acacia boormanii is equally showy in late winter. It has a long linear leaf.

Acacia pravissima on the top of the viewing escarpment.

Acacia pravissima with its unique triangular phyllodes along the stem instead of leaves.

Acacia mearnsii has distinctive creamy-white flowers.

Fortunately, we had not limited our trees to *Acacia*, although they had made up the bulk. What other trees could we source that had a leaf resembling *Acacia* or looked a little different and would only be used in the savanna? An obvious one was false acacia or black locust *Robinia pseudoacacia*, native to the southeast of the United States. It does look like an *Acacia* even if it is not, hence the common name. It is often a pioneer species enriching the soil as it goes, growing vigorously, maybe 2m a year (that legume family beneficial symbiosis with bacteria again). It seeds well and suckers a lot, and can often take over grassland by shading it out and outcompeting. It is so good that it is invasive out of its natural range – even in parts of the United States. *Robinia* seen growing in the French Pyrenees were very popular with honeybees but already made up 10% of the forest – what would that be like in a hundred years? The trees prefer dry or well-drained conditions. The sides of railway tracks have become a favourite spot too. But it needs hot summers to ripen the wood. Most Irish summers are too short, chilly and damp, meaning the wood stays brittle and shoots break off on windy winter days. In suitable climates, *Robinia* will equal an oak in size but very seldom in Ireland. Planted in groups around the savanna, many died, especially once shaded a little. They grew quickly as small trees – so far, anyway. All young shoots, especially suckers, are viciously thorny, no doubt as defence against animal browsing, but it worked equally as well against staff.

Three large root-balled American honey locust *Gleditsia triacanthos* 'Shademaster' (a nursery selection of the species) were planted at the very top of the raised visitor path to add a mature aspect but with not much hope of them doing well. Unfortunately, that is exactly what they did – they hardly grew at all. Climate would be to blame again, too cold and too short a summer. This tree is native to central United States, so it is used to extreme heat. Equally tolerant of poor soil as the *Robinia*, the species can be a seriously invasive weed, especially in Australia. Thorns are a real hazard – even puncturing car tyres. Branches will have thorn clusters up to 200mm long, with each thorn being branched – three thorns, hence *triacanthos*. Ferocious! While visiting Zurich Zoo, Budapest Zoo, and Berlin Zoo over the years during zoo horticultural conferences, *Robinia* and *Gleditsia* were seen as street trees. They tolerated the roadside conditions and typical central European hot continental summers well, ensuring they grew really well. But with so little frost in Dublin, loads of plants could grow outside all year that would not survive in Central Europe. Touché. The other man's grass is always greener. A few small Caspian or Persian locust *Gleditsia caspica*, grown from seed in the zoo nursery, grew very slowly and had equally nasty spines on them.

Pride of India or golden rain tree *Koelreuteria paniculata*, is aptly named as the flowers fall to carpet the ground in yellow. This is obviously a wonderful tree. It has large pinnate leaves of up to 450mm and a great display of yellow flowers followed by unusual papery seed pods. There is a great specimen in Trinity College for anyone in Dublin, overhanging the boundary wall halfway along Nassau Street. Native to North China, Korea, and Japan, this has travelled around the world as it is such a good landscape tree. Flowering starts when still young, even at 2-3m tall. At that height, you can really appreciate the flowers with a reddish hint to the lower part of the petals. In spring, new leaves are a bright pink. Very eye-catching. There's autumnal colour with a good yellow finish. This is now included in Sapindaceae, the Soapberry family, along with horse chestnut *Aesculus* and maples *Acer*. A few specimens were planted from various nurseries, and they are slightly different in flower and flowering time. Maybe they simply had a different seed source. Interesting.

Koelreuteria paniculata is a glorious show of colour in late summer. It even flowers well as a small tree. These are on a tree no more than 3m tall.

From the United States and Mexico, bastard indigo *Amorpha fruticosa* looks the part. It has a similar leaf to *Koelreuteria* but is only 100mm long. It has an interesting flower, though you almost need to be told it is in flower to notice it. Short, vertical racemes of purple flowers with contrasting yellow stamens are produced each summer and do look good, but each flower is only a few millimetres long. Growth is too vigorous in Ireland due to richer soil and wetter summers. As for *Robinia*, with a cool short summer, the wood is not ripened. Winter winds usually break the current years' growth to the point when pruning it back in early winter is best, rather than having breaks and damage where unwanted. In Burgers Zoo in Arnhem, Netherlands, the same plant was used in their savanna area. Great minds think alike. Beware – it is one of the most invasive plants in Europe. If you walk along the River Tiber a few miles north of Rome in Italy, as some of the delegates attending the horticultural conference at Rome Zoo did – you will find it popping up in places you don't expect it. The common name is a reminder that it has been used as a source of blue dye.

Large shrubs were needed in places near the visitor pathway. Something different looking and savanna-style was needed for the pathways in between, and especially to screen the very necessary heavy gates for the rhinos. One plant in stock from our own propagation efforts in the nursery was *Olearia virgata* var. *lineata*, a shrubby daisy, but that description does not do it justice. It grows vigorously to 4m or so, with long drooping shoots and incredibly thin leaves (like rosemary), which are dull green on top and silvery underneath. The flowers in early summer are really small, a few millimetres, but en masse they are noticeable, especially if downwind on a good day as they have a subtle perfume of maybe vanilla, it's hard to call. Despite the thin leaves and twiggy growth, this has formed a dense canopy, completely screening part of the rhino habitat as visitors walk down from the escarpment. This was probably originally planted in the older part of the zoo near the

Amorpha fruticosa, an intriguing flower, but only when really close as it is so small.

red pandas by Robert Lloyd Praeger, the Irish naturalist, while he was on the zoo council.

We had one seedling of weeping broom *Carmichaelia stevensonii* (*Chordospartium* in old money) in our nursery. Sown from seed in 2005, it is a most unusual plant with thin narrow leaves – more twig-like and pendulous. The pea-like flowers in summer are technically white but with purple markings that make them look more pale mauve. Eventually, it can reach 5m tall. One ancient one seen in the Logan Botanic Garden in Scotland was a proper rain shelter in a gentle shower. Wonderful. Unfortunately, it is under threat from habitat loss in its native New Zealand. It has been seen for sale very occasionally in more recent years.

Smaller shrubs were useful too. Even if they were only 1.5m tall, they would be a useful screen, especially if the planting area had any rising slope. One of the most useful has been spider flower *Grevillea rosmarinifolia* 'Canberra Gem,' a vigorous evergreen with rosemary-like leaves, but very sharply pointed. The flowers, produced from late winter to early summer, are very strange. They have no petals, just an elongated calyx tube that is a good red colour, with an even more elongated style, thus the common name – not to be confused with *Cleome* or spider plant *Chlorophytum*. Common names can often complicate matters. Regular pruning back keeps it in check, but it is not easy to prune 'nicely,' though gaps do fill in again quickly. Hard frosts can cause damage, but more often than not, just a section of the plant is damaged, and pruning out allows regrowth. *Grevillea* is a member of the Proteaceae, which includes the incredible *Banksia* and *Hakea*. *Grevillea* itself is a most diverse genus, from low shrubs to tall trees. Recent taxonomy research has Proteaceae related to Platanaceae – well known for plane trees and *Nelumbonaceae* – very well known for the water-lily-like *Nelumbo*.

Grevillea rosmarinifolia 'Canberra Gem,' is a great show in early spring.

Another soft, spikey leaf shrub was blue pea *Psoralea pinnata*, with its most wonderful dark blue flowers. This had grown well for several years near the original reptile house and was an African plant too. Unfortunately, not hardy enough as a heavy frost killed it off and any other *Psoralea* spp. tried have not been as good. If you see it, buy it, try it – you will love it while it survives. The Latin name comes from *psoraleos*, meaning scabby as there are glandular dots over the leaves. Maybe the common name should be scabby pea, but then it might not sell as well.

Psoralea pinnata.

Another useful spikey leaved shrub is *Leptospermum*, but many plants of *L. scoparium* 'Winter Cheer' and 'Red Damask' were killed outright in the bad winters of 2009/10. Replacements and new cultivars such as 'Silver Sheen' will probably get through the average Dublin winter, but a different species *Leptospermum namadgiensis*, promises to be hardier. It is very new in trade, so time will tell. From seed collected high in the Australian Alps in the Namadgi National Park, it has silvery leaves, which makes it very distinctive, and it has masses of white flowers in early summer. *Leptospermum* are very well known under the common name tea tree. Absolutely nothing to do with standard tea (which is made with young leaves of *Camellia sinensis*), but Captain Cook tried the leaves to make tea and found it ok, so the name stuck. It is better known for its medicinal properties, under the New Zealand Maori name *manuka*, and as the source of excellent honey too.

Bladder senna *Colutea arborescens* is another unusual shrub, and it grows to about 3m but is not too dense. It has yellow pea flowers and a most unusual seed pod that is 60mm long and inflated like a small balloon, hence its common name, though it is not related to *Senna*. Used as a soil stabiliser, it has in some places become a nuisance – yet another introduced alien. This was familiar to me from botanising near home in London years before where it had come in with bomb damage rubble, no doubt, and it likes good drainage, so it was well suited to it.

There was a good diverse appearance with enough small trees and shrubs, some in large numbers and others dotted around, so what about the spaces in between? Grasses, herbaceous, and bulbs? So much to choose from and there is a very interesting range of real African plants that are often grown in our gardens and are readily available.

Grasses had to be a big part. Our most important consideration here was for a twelve-month season. Some ornamental grasses die down in autumn and do not surface until the following May – a long six months of bare soil. Careful placement could hide this but consider the work as well. If too many needed cutting back, and all by hand, it could be an issue. There was always the option of grassy-looking plants, that the average visitor may not even think about but would see just a green background and that seemed a better option, with less work.

New Zealand wind grass *Anemanthele* (*Stipa*) *lessoniana*, with a plethora of other common names such as pheasant's tail and gossamer grass, was one of the best. The flower stems move in any wind, gently nodding and swaying, hence the various names. It grows easily from seed, and most likely a couple of thousand 40mm plugs (very small plants) were planted. When used near the Meerkat Restaurant, the same plant had done extremely well, growing to half a metre, with a bit more when flowering. It has good green leaves in bulk, which stay greener if in the shade, but often turns a reddish colour in full sun. Flowers are borne in profusion with a pinkish brown hue, but it is a typical grass with very small flowers. Deadheading the flower heads is very easy if done at the right time – just as the seeds are ripe because the whole stem pulls out easily. If not deadheaded, they can collapse onto the ground and be far less decorative, but far worse, they spread their seeds, which can be a problem. Seedlings began to pop up in the most challenging growing conditions – the edges of the path, around manholes and drains, between cobbles or bricks on pathways, but very, very seldom in a planted bed with good soil and well mulched. Very odd indeed. It obviously prefers a well-drained, challenging spot. Each plant is also short-lived. The leaves associated with each flowering stem dieback, giving a brown appearance. After a few years, the whole plant may die but then persist for a year or three, looking more dead than alive. As a garden plant, those dead leaves and the half-dead plant would be not decorative at all and would be pruned out or cut back to regrow perhaps, but for the savanna, they were the best part, giving the desired dry savanna look all year and from a distance too. It is not often that a dead plant is better than a live one. Indeed, zoos are strange places for gardeners to work at times – maybe zoo-illogical...

Anemanthele lessoniana gives a natural grassy background to the top of the sandstone rocks from any angle. Purple *Delosperma cooperi*, silvery *Lotus* (*Dorycnium*) *hirsutus* (or the hairy Canary clover!) and yellow *Aloiampelos* (*Aloe*) *striatula* give interest near to the visitors.

Even when the dead *Anemanthele* were removed as they finally collapsed, there was often a use for them. A small breeding colony of endangered waldrapp or northern bald ibis *Geronticus eremita* had nests high on the ledges of the artificial cliffs in their habitat at the lower end of the zoo. Theirs had been the first habitat created taking inspiration from the wild for the design and in the wild, they indeed breed on cliff faces in dry desert areas with a stony substrate. Every chick bred in captivity is important, especially as they are endangered in the wild. The ground surface in their zoo habitat was mainly 25mm round

Red *Crocosmia* 'Lucifer' and white *Dierama* 'Guinevere,' cultivars of African native plants. Note the *Rubus* 'Betty Ashburner' on the gabions in the distance, completely hiding the stones. This picture was taken two years after planting and before the rhinos found the *Rubus* to be a tasty snack.

The savanna theme needed succulent plants, and there is no better option than the hardy ice plant *Delosperma cooperi*, with its small fleshy leaves and flowers that look like the annual summer bedding plant *Mesembryanthemum*, which they are related to, but with the great advantage of being perennial. Preferring really dry spots, like the *Crassula* above, *Delosperma* enjoys the path edges as you enter the savanna, with great drainage on top of the gabion and sandstone retaining walls. With a mere trowel of topsoil on top of a large sandstone block, one plant survived for many years, growing in almost nothing. Amazingly, for such a succulent plant, it is completely hardy. A clever adaptation in many species of this family is the hygrochastic seed capsules that sense the humidity in the air and only open when it is the best time for the seeds to germinate.

Masked twinspur *Diascia personata* has a long flowering season and is an excellent plant with 15mm pink flowers all summer. A stunning plant, though vigorous growth to 1m, needs regular pruning back. The best way to treat *Diascia* is as a temporary planting as it is short-lived. Cuttings are so easy to root as spares, allowing planting wherever needed. It hails from the East Cape of South Africa.

Some plants are real icons. The angel's fishing rod or wand flower *Dierama pulcherrimum* is one. From an evergreen grassy-looking clump (handy to have for that alone) come long, thin arching flower stems each summer of up to 1.5m, with pendulous flowers in white, pink, and even a very dark purple that looks almost black. The weight of the flower on such a thin stem makes it move in the gentlest of breezes, hence the name. There are many species of varying heights and colours, and very many cultivars too. Several were planted. Seeds are formed readily and drop off nearby to start a new plant, often where it is not wanted! Deadhead religiously after flowering. Be aware that they flower from the top down, so the top seeds are ripe first. Native to a wide area of Southern Africa. The name is from the Greek *dierama* meaning a funnel, which is the shape of the flower, while *pulcherrimum* means very pretty.

Pineapple lily *Eucomis*. Here we go yet again – this is not a lily at all but actually in the asparagus family. Nor is it in anyway related to pineapples! It has to be one of the most asked about plants when in flower. Each flower spike has a wee clump of leaves on top – exactly like a pineapple, hence the name. Native to many countries in Southern Africa, there are several species and several good cultivars too. Some of the best in the savanna have been *E. comosa* 'Sparkling Burgundy' and *E. bicolor*. Both are good solid growers, even self-seeding in some spots. Summer rain, drier winter, and not too cold a winter would be their preference. They fit into the standard zoo system of regular mulching and no lifting for winter. That may not suit all species or cultivars.

There are several other real beauties around. *Eucomis* has one fascinating property, especially for a bulb. It can be propagated by leaf cuttings or the wee pineapple growth on top of the flower. Simply divide up the leaf into 50mm sections or use the top growth intact and bury the bottom edge in pure grit and keep it just moist, not wet. The leaf will largely shrivel up and brown, then the energy of the leaf will be passed down to a few small bulbils that form on the basal cut edge. These will grow the next summer. Very handy. But why does it do this? Imagine the natural habitat, often rocky spots, with the bulbs growing in organic debris as epiphytes. Any broken-off leaves would form bulbils that would be an extra chance of keeping the plant going and maybe spread. Belt and braces for species continuing. One other aspect of the flower on some species is the scent – like an incontinent tomcat has passed by. A colleague regularly threw a neighbours' innocent cat off their patio, blaming it for the aroma…

Eucomis comosa 'Sparkling Burgundy,' excellent colour even if only in leaf.

The more flowers we could pack into the savanna, the better. *Euryops* fitted the purpose well. It is mainly a genus of small shrubs in the daisy family, with about 100 species, mainly from Southern Africa but a few elsewhere in Africa. There is only a handful in commercial cultivation in Europe. Most often seen is *E. pectinatus* with downy, silvery-grey leaves that looked particularly apt in the savanna. Forming a shrub up to 2m tall, with good yellow daisy flowers 25mm across almost all year, it is making a good screen, but it is not fully hardy, tolerating only a couple of degrees of frost at most. Fortunately, it is incredibly easy to propagate from cuttings, so keep a few spares in the nursery greenhouse and risk as many as you like once you have that insurance. *E. chrysanthemoides* is almost identical, except the leaves are plain green, and it is not as hardy, nor did it look as good as a savanna plant. *E. tysonii* turned out to be a real gem. It has a typical yellow flower of up to 15mm across, held at the end of each shoot. This is a sprawling sub-shrub that clambers around rocks with the most natural effect. It is not as good or as showy at flowering but is very nice, far hardier, and again, very easy from cuttings.

Yellow *Euryops tysonii* scrambles in a very natural way across the sandstone.

Bulbous plants were an obvious choice, though bulbs in an open garden situation can be a problem once they die down. Where they are exactly, and if they don't like the standard mulching regime, they may not thrive. Summer hyacinth *Galtonia candicans* is reliable. They produce pendulous pure white flowers each summer, growing to about 450mm high, and *G. viridiflora* with greener flowers are nearly as good. This is another member of the asparagus family.

Better known is the genus *Gladiolus*, often seen in gardens as a multi-coloured flower mix 1.5m tall in lime green, yellow, orange, pink, red, blue, white, and everything in between – very colourful indeed, but not the right sort of plant for the savanna. Far too gardenesque. Of the several species tried, the best were *G. tristis*, *G. oppositiflorus*, and *G. dalenii*.

G. tristis with its thin grass-like leaves flowers early, normally in April, (coming from a winter rainfall area). *G. oppositiflorus* is a great flowery show of mainly pink. *G. dalenii* with an almost orchid-like flower is a most unusual species with several races (botanically a recognition of regular differences that do not quite make a sub-species) that, uniquely among this genus, have diploid, tetraploid and hexaploid genetics. This is why there are so many *Gladiolus* garden hybrids; they hybridise so easily. There is still plenty of scope for hybridising the smaller *Gladiolus* species and getting some choice, hardy, delicate colours on smaller plants than the usual garden hybrids. Unfortunately, a minor rabbit invasion caused much damage in the savanna, especially to new spring growth, and flowering was lost for a year or two. The name is from the Latin *gladius* for a Roman legionary sword; *gladiolus* means a small sword, after the leaf shape.

Gladiolus tristis, a reliable spring flower with grass-like leaves.

Gomphrostigma virgatum is another silvery foliage small shrub, with small white flowers at the end of each shoot in summer. It tolerates pruning if too big. Preferring moist soil, it grows like willows in Ireland on the edge of streams, so much so that one local South African name for it is otter bush – they must enjoy making their holts under the mass of stem. The well-known shrub *Buddleja* is a close relative.

Another iconic plant everyone knows is the red hot poker *Kniphofia*, the common name taken from the red or yellow flowers on top of a solidly erect stem. Kniphofias are found throughout Africa. The genus was named for Johann Hieronymus Kniphof, a German physician and botanist in the 18th century, who collected many plants and formed an extensive herbarium (dried, pressed and mounted specimens). Several species and cultivars were planted throughout the savanna; a few came in as ex-plant show pots (as did some *Eucomis*) from a nursery. Old well-grown plants make an immediate impression, but unfortunately, they had no labels. Others suffered rapid planting and lost labels in the process – the names are there somewhere!

Really good plants include *K. thomsonii*, an unusual one, with widely-spaced pendulous flowers of orange and yellow. Individually, they resemble *Phygelius*. Very dignified. The roots run a little bit to make a good clump. If not too cold in the winter, it will keep growing, throwing up flowers at any time. Uniquely architectural is *K. caulescens*, with a stout stem of 50mm thick that creeps over the ground, rooting as they go. The glaucous blue-green leaves are evergreen but die off a dismal brown and are best removed – which shows the stem better anyway. Its stunning flowers on rigid 1.5m stems are red in bud but open to a yellow flower, from bottom to top, giving a two-colour display. Wonderful, but there are some questions as to identification, as it may be a hybrid.

Another equally architectural species is *K. northiae*, with very wide, long leaves. The flower is a massive poker; it stands at 1m on a stout 50mm stem with orange flowers when in bud, then yellow when open. The whole flower head is far larger than normal. It is a very impressive plant, but the leaves do need room to sprawl. Originally planted in the older part of the zoo in a less sunny spot, the plant had died down after flowering, and when dug up, it was a stinking mess of too-wet rotting roots. But life was still there, and a few 'daughters' were teased out, which were big enough for the savanna in a year or two.

Another really weird-looking species is *K. typhoides*. The specific epithet means 'looking like a *Typha*' or bulrush, and the flower does exactly that – a 20mm wide 200mm long brown poker. Each small brown flower has small yellow stamens. You would need to know it was flowering to go and look at it. Even then, you would wonder, but interesting, nonetheless! One excellent cultivar for sure is *K.* 'Wrexham Butterfly,' with bright, pure yellow pokers, very eye-catching indeed.

Yet another iconic plant is *Melianthus major*, honey flower from the Greek *meli* meaning honey (as in *Apis mellifera* honeybee). The beautiful large blue-green leaves are architectural, held on stems of up to 2m in a good year, though a really hard winter will take them all to ground level only to regrow strongly. The flowers in spring and summer are borne on spikes (technically a raceme) at the end of each stem, with several dozen flowers on every single one. These are usually a dark red to mauve colour, but they can be green on some plants – perhaps a different clone? The flowers are unusual in that they move through 180 degrees, being vertical in bud and in flower, but twist around and hang down after pollination.

The nectar is freely produced to the point of dripping; even more unusual, it is black. It is very attractive in its natural habitat to sunbirds and other nectar-feeders and bees wherever it grows. According to some sources, the nectar is the only part of

Moraea huttonii, seldom seen but really worth growing.

Other plants used as deemed to have the right appearance for a savanna, looked different enough, were interesting, or had a growth habit that was useful, but were not from Africa. These included the following:

Broad-leafed glaucous spurge *Euphorbia myrsinites*, with its blue-green spirally arranged glaucous leaves on a sprawling stem, is a useful evergreen and is very drought tolerant. As with the *Geranium* above, *Euphorbia* has a great seed dispersal system. There are normally three seeds per pod, and each pod splits apart with great force when fully ripe. Collecting seed is a tad iffy as the sap is thick, white, and sticky, and not nice on bare skin. It dries black and glue-like, so the less handling, the better. Cut off the flower head when nearly ripe, place into a labelled paper bag – raiding the front gate for paper carrier bags became part of the routine in summer for all seed collecting – close over and wait. Now, if you place the bag where it is nice and warm, as the seeds are ejected with pieces of pod, there will be an irregular – *ping, tinkle, tinkle*. Doing this at home (don't ask), one colleague had a very worried wife convinced there were mice in the room, but there was no sign of them...

Bird of paradise sisyrinchium *Sisyrinchium palmifolium* is a small, typical yellow *Iris* family flower of 15mm, with a flowerhead a little reminiscent of a bird of paradise or the other plant with that name *Strelitzia*. Usually, there are dozens of flowers on each 450mm stem, each may last only a day, but with so many in succession, there is a longer period of colour. A good show. The flowering shoot then dies and needs cutting back; a little bit fiddly and tedious amongst the remaining evergreen shoots.

Bluestem or Ma Huang (it comes from Asia…) *Ephedera equisetina* – you either love it or loath it. This is an evergreen sprawling, twiggy bush with no leaves. A survivor with fossil records of similar species going back 150 million years. Plants are either male or female (dioecious), so you need both to get seed though it can be propagated from the gently spreading roots. It looks similar to the awful weed horsetail *Equisetum*, hence the specific epithet. A real novelty to grow, a talking point with the right appearance for the savanna but is not a colourful garden plant at all. Love it.

Iris wattii and *Iris confusa* are both called bamboo iris, with wide shiny leaves on 450mm stems, topped with orchid-like, unusual, and reliable flowers. Both deserve to be planted more.

Several different species of New Zealand satin flower *Libertia* were used here as grass lookalikes, tough and reliable. There are far more details in a separate chapter on this indispensable plant.

Libertia chilensis Formosa Group – what else would survive in such a busy viewing area, with kids running through it all day.

Paradisea lusitanica is a vigorous bulbous plant. White flowers are borne on 1.5m stems in early summer, poking up through other vegetation. Be warned, it can self-seed very easily. Seed would normally be collected for distribution through garden society seed lists. One busy year while planting up the gorilla habitat, that simple quick task was missed, and there was a verdant lawn of seedlings around it the next spring.

The bayonet plant *Aciphylla squarrosa* was grown from seed in the zoo nursery. It is most unusual looking as it is basically all spikes, which the common name suggests. The Latin *acicula* means needle, while *phylum* is Greek for leaf. It is most certainly a well-armed plant. Warning: approach with caution from any angle at all times. The 150mm needles are rigid, very sharply pointed, and like a hedgehog, point every which way. The greenish-yellow flowers are borne in a 450mm long umbel and are armed with extra needles. Few people would guess that they are looking at a member of the carrot family Apiaceae (or Umbelliferae). Unique and covetable.

Aciphylla squarrosa.

Asphodel *Asphodeline* is from around the Mediterranean region. Good reliable performers, herbaceous, and form clumps of vertical stem, clothed with narrow leaves. King's spear *A. lutea* to 1.5m is easily available. Its six-petalled yellow flowers on top of the vertical stems are showy and great for pollinators. More graceful and delicate looking is Jacob's rod *A. liburnica*. A little later, shorter and smaller in all parts, it is really lovely with a delicate appearance, but less often seen. Pollinated by moths, the flowers open late in the afternoon.

Asphodelus is very closely related, and you can have many interesting moments looking up *asphodel* for these too. Enjoy. There are several species around in both genera, often with the wrong names attached. *A. aestivus* is excellent with white flowers in spring on 1m stems. *A. albus* is similar, but the leaves are glaucous.

Nippon daisy *Nipponanthemum nipponicum* was only planted as it has a typical daisy-type flower, but what a gem it turned out to be. The name is a bit of a giveaway that it hails from coastal Japan – about as far from the African Savanna as you can get, but it is unusual in cultivation. It is a shrubby chrysanthemum. However, now it is in its own genus and the only species in it. The large leaves are evergreen, though they suffer if too cold, and the shoots die back. Light pruning is needed once all frost is past. White daisies on steroids, flowers of up to 9cm cover the 1m high bush. It is a welcome dome for many weeks in autumn, spreading the flowering season for the savanna. It is also found in trade as the Montauk daisy, named after the New York area, as it has become naturalised on the coast there and in New Jersey.

Nipponanthemum nipponicum starting to flower – in November.

Tiger flower *Tigridia pavonia*. Alright, this is another completely non-savanna plant, but it is a useful bulb with spectacular flowers in red and yellow. It has wonderful patterns at the base of the petals inside, hence the common name. Originally from Mexico (Tigers in Mexico?? Common names again…), it self-seeds a little, and pops up in odd spots. It gives great splashes of colour for a very short time, as each flower lasts only a day.

Anchor plant *Colletia paradoxa* is yet another equally non-African Savanna plant, hailing from Chile and Argentina. Eventually, a large shrub growing to 4m after many, many slow years. It should be approached with great caution; like the *Aciphylla* above, as it is phenomenally well-armed. It has no leaves; instead, it has wide spines (technically cladodes) shaped a little like an anchor (hence the common name), which act as leaves. The spines stick out at every conceivable angle, sharp and very rigid. Small white flowers appear in late summer and although only a few millimetres across, they emit a sweet scent that can be elusive to track down unless you know it. You can risk getting up close and personal to enjoy the aroma, but you may want to insure your nose first. Seed is set most years, but the seeds are attached at the base of those spines. It is only the second plant you'll ever need wide-nosed pliers to collect seed from, carefully poking them in amongst the spines.

The other plant you will need pliers for is *Onopordum acanthium*, a giant thistle. This is planted as a temporary filler occasionally. The pliers come in handy to pull the *Onopordum* seed heads apart, but do it outside because the hairs around the seed are not nice if inhaled. *Colletia* is in the buckthorn Rhamnaceae family and as with many of that family, it has nitrogen-fixing nodules on the roots, as found in legumes.

Colletia paradoxa, each flower is only a few millimetres across.

Ichang bramble *Rubus ichangensis* is a wonderful blackberry from China, with long panicles of very small white flowers. The bees and other pollinators love it. The whole plant is exceptionally spiny but with tiny, hooked thorns. Once it gets you, you're got for sure. Used to cover a steep bank, a small number of plants quickly filled the space and layered in well too. It is a typical blackberry, an empire builder. Very handy.

Necklace vine *Muehlenbeckia complexa* from New Zealand is a dock *Rumex* relative and does indeed grow like a weed. Incredibly subtle, it spreads its thin, wiry stem along the ground, up fencelines, or other plants. Often not seen until it has a real presence. It will get to the top of a 4m hedge or wall-trained plant in a year without bringing any attention to itself. Only then will you notice the whole top of the plant is covered with this thug, which has alternate common names such as mattress plant or wire vine – both very apt. Great for clothing a fence or shed to hide it, in a very informal way, but be aware that it will have empire-building desires given half a chance. It was used on a few fences along the savanna trail purely to screen and soften the fence line, giving soft growth, gentle on visitors' eyes.

In a few spots, we needed planted screens and had to think about what plant would be best or would even survive. Hiding the animal houses was essential, but with giraffes and zebras able to browse a little bit, the planted screens would need to be edible just in case, plus resilient enough to grow back. The best or rather the worst, as it was not planned, view of the animal houses was from the raised walkway, which meant we needed a tall screen, trees not just shrubs. This meant a trawl through what was available in good sizes, not a poison risk, and would grow back again. Walking through one tree nursery, the grower commented that he hated *Tilia x europaea* 'Pallida' as it always reshoots from the main stem when he tries to grow a nice clean leg. Now, there's a thought... This is a growth habit with *T. x europaea*, and often, the entire main stem on old trees is completely hidden in scruffy twiggy growth, which is removable, but it always comes back. It is a tough tree, cheap, easily grown, used a lot for large estate avenues, and it always grows back! Right, that's sorted so.

To make it just a bit more challenging, the trees had to be planted within the line of the rhino-proof fence. Theis fence was constructed from large steel upright posts in concrete foundations 1m wide and 3m long, plus the posts were 2m apart. That meant the root-ball had to fit between two long concrete strips, 1m in depth. After a few gentle discussions, the foundations were lowered, which would allow the tree roots to grow over the concrete in about 300mm of topsoil. Planting needed mini diggers to lift the trees in, so the builders needed to be on our side. On the day, the root-balls sat balanced on the concrete foundations, topsoil was carefully teased all

Crocosmia 'Lucifer' and *Euryops pectinatus* frame the view well.

Buddleja colvilei, with each flower to about 25mm, makes a stunning display within planting that separates the savanna from the Gorilla Rainforest.

Echium wildpretii took three years from seed to flower and needed some light conifer branches over it to protect it in bad frost, but it was certainly well worth the effort. Unfortunately, it dies after flowering.

Opening Day

On opening day, the President of Ireland was due in. It was the last morning, the last two hours of work – was it all ready? Well, yes, but only at the very last minute. The builders were trying to do the last fence and last path edge, the horticulture team put the last mulch in place and spruced the planting a little. Meanwhile, two large water tankers were in use, washing down the roads with power hoses, everyone was tripping over one another, with at least twelve vehicles on the road forty-five minutes before opening, the podium, speaker cables and chairs were being put in place – it was definitely last minute and as breathless as this sentence. It was much more hectic than the Kaziranga Forest Trail last day, for sure.

were established in odd spots around the zoo in blue and white form, probably from topsoil movements years before. The intention was to keep the native species and have no hybrids, keep the gene pool pure.

Lords and ladies or cuckoo pint *Arum maculatum* is another very early flower but is most unusual in appearance. As with all aroids (arum family members), a cowl arches over a small finger-shaped spadix. Red berries follow in autumn, but beware, all parts of this plant are poisonous. An excellent educational plant. Many aroids give off heat from the central spadix, which warm any insect pollinators, and causes air currents to rise, taking any scent produced with it. Aroid flower scent is often unpleasant, resembling rotten meat to attract flies, the main pollinators. The leaves are sometimes spotted with dark purple patches, hence *maculatum*, but most plants in Ireland are plain, dark shining green.

Primrose *Primula vulgaris*, which everyone knows is a sign of spring, was tucked under the hedge in places, pushing its ever-welcome bright yellow flowers out to cheer everyone. It is also great for pollinators. Here *vulgaris* means commonly found, which it is, but still very welcome.

Teasel *Dipsacus fullonum*, a biennial, was planted too. A rosette of leaves forms in the first year, followed by a 2m flowering shoot in the next year. It is a prickly plant with a great architectural presence even in winter as the dead stem lasts well. The blue flowers are great for pollinators and open artistically, starting in the middle of the cone-shaped flower head, opening in two circles up and down, slowing getting further apart. The leaves clasp the stem very tightly, allowing water to collect, great for insects or birds to drink. The generic name comes from Greek *dipsa* for thirst. Better yet, the seed heads are very popular with seed-eating birds in winter, especially finches. It can seed about a bit but is easily removed. The common name relates to teasing, as does *fullonum* which relates to fullers' (wool workers) use of the dried seed heads to stroke the cloth to raise the nap on the cloth, making a softer surface.

Harebell *Campanula rotundifolia* is very colourful tucked in under the hedge. The bright bluebells bob in the slightest wind but almost disappear amongst taller growth around it – regular editing of nearby thugs would be needed. The Latin *campana* meaning a bell, is obviously for the flower and can be seen (and heard) in bell ringing – campanology. The round leaves are *rotundifolia*.

Knapweed *Centaurea nigra* is a common plant but very welcome. The flowers are always busy as they are one of the best nectar sources for bees and hoverflies.

Hemp agrimony *Eupatorium cannabinum* is a tall perennial for the back of the meadow. Flowers appear in late summer – it is good to keep the nectar sources spread over the season. Butterflies love this. The small flowers are in wide masses and are a good landing spot. Don't mind the name *cannabinum*; it merely means that it resembles cannabis, having none of the herbal 'properties.' The generic name is the Greek name for the plant and commemorates Mithridates VI Eupator, King of Pontus (much of modern-day Turkey, Armenia and coastal states of the Black Sea), 132-63B.C. *Eupator* meaning good or noble father and was used for several rulers. How is that for an education lesson from a plant?

Meadowsweet *Filipendula ulmaria*, with its mass of creamy flowers in summer, is a common but lovely plant. It is often seen alongside drainage ditches on country roads. This was the first source of aspirin in the 19th century when salicylic acid – as discussed in willow *Salix* – was isolated and used as a disinfectant and painkiller. A great story for the education team, especially if linked to the gorillas and willows. The epithet *ulmaria* refers to the leaves which resemble elm *Ulmus*.

Meadow cranesbill *Geranium pratense* with intense blue flowers is great for early summer. Each plant forms a dense, round clump a metre tall. A fantastic show and it is excellent for pollinators. *Geranium* is from the Greek *geranos*, meaning crane from the long, beak-shaped structure of the seed head. Other related plants are also called cranesbill, while *pratense* simply means 'found in meadows.'

Ground ivy *Glechoma hederacea* came in, as did several of these meadow plants as cuttings rescued from hedge maintenance were noticed on walks in the country. Planted in the dense shade at the foot of the hedge, it rapidly spread – really romping across the good soil in the meadow and needing control. Well named, it looks like ivy, hence *hederacea*, but it does not climb. Popular with bees, despite the small size, it is a good nectar source, as are all members of this, the mint family. The rampant growth, especially in good soil, has made it a real problem in parts of North America, where it is an alien invasive plant and very difficult to control as it roots from each node along the stem as it grows.

Stinking hellebore *Helleborus foetidus* is not native but is sometimes found as a garden escape – another good talking point for educational tours. It is a good bee plant very early in the year but often badly infested with greenfly – food for birds. The seeds spread far too easily, but simple deadheading after flowering stops this and keeps it looking tidy. Stinking – yes. Indeed, the leaves are unpleasantly pungent, hence *foetidus*.

Field scabious *Knautia arvensis* is yet another great pollinator plant. It is in the same family as teasel mentioned above with nice blue flowers.

Ox-eye daisy *Leucanthemum vulgare* can be a real thug, taking over if allowed, but it gives a great show of white flowers in summer. The common name is from 'day's eye,' as the flowers open each morning and are quite large, hence ox-eye. *Leucanthemum* is from the Greek *leukos* meaning white, and *anthemon* meaning flower, while *vulgare* is simply common.

Ragged robin *Silene* (*Lychnis*) *flos-cuculi* is a real charmer, with split petals – hence the ragged name – in a delicate pink shade, not exactly robin colour. It prefers damp soil, though. The specific epithet is from the Latin *flos*, meaning flower and *cuculus*, meaning cuckoo, as flowers appear from early in summer.

Silene flos-cuculi has the most attractive delicate flowers, here flanked by ox-eye daisy.

Purple loosestrife *Lythrum salicaria* can be a little thuggish, growing to 1m. It is another damp-soil plant often seen making drainage ditches on the roadside very colourful when in flower, but it grows happily in drier conditions too. An excellent pollinator plant, and a great story for the education team here. This plant is a very invasive alien species in New Zealand and parts of North America, significantly decreasing the natural biodiversity. Seeds produced in the thousands spread very easily. The problem was so bad that biological controls have been used, importing a species of leaf beetle, *Galerrucella calmariensis*, native to Europe and Asia, that strips the leaves of the plants. However, beekeepers love it as the flowers produce plenty of nectar for their bees.

Cowslip *Primula veris* is a spring beauty, popping out from under the hedge before it comes into leaf. The flowers are a lovely bright, deep yellow. For the educators, it has a good story too. Well, maybe – why is it called 'cowslip'? There used to be so many cowslips growing in a meadow that the farmers said they grew wherever a cowpat or slip had been – something kids will remember for sure.

Red campion *Silene dioica* gives a great show for a long season, with flowers of a more pink than red hue. Very colourful, but far too happy to seed around if any gaps.

Red campion two months after planting. With too little deadheading, it seeded too much and took over.

Marsh woundwort *Stachys palustris* is often seen but seldom looked at. It has flowers that are not showy but very good for pollinators, as are all members of the mint family – Labiateae. Strongly scented though not pleasantly, again, this is common in this family. Typically prefers damper soil, so may not do well here – time will tell.

Hedge woundwort *Stachys sylvatica* is very similar to *S. palustris* above but more strongly scented. It is tolerant of drier soil, so it has a better chance of doing well.

Devil's bit scabious *Succisa pratensis* is another great pollinator plant with a long flowering season and good blue flowers. Yet another member of the teasel family. For the educators, this is an interesting plant as it is the main food plant for the larvae of the marsh fritillary butterfly *Euphydryas aurinia*. Devil's bit scabious? Well, the root is cut off sharply, truncated – as if bitten off from below – only the devil himself could do that. Hence the genus name *Succisa* from the Latin *succido*, meaning to cut off.

The first year, the meadow looked great with plenty of colour, and both hedges were low enough or thin enough to see through or over easily. A thick mulch of tree chippings – without manure as no feed was needed or wanted – kept weeds to a minimum. Deadheading was done as required, along with a little weeding, but those creeping buttercups were starting to get too well established. With more seeds germinating, it was going to be difficult to get out of a large dense clump of any other plant, plus there was always other work needing attention. The second year saw many more seedlings coming up within the mulch. This was mainly from the red campion and ox-eye daisy, and while the latter was easy enough to weed out, the former was in very dense masses. With a long flowering season, deadheading had probably been too late, or perhaps it had even helped to spread the seeds as the arisings were removed by walking the length of the meadow to the only gate at the one end.

The two long hedges enclosing the meadow were very different, despite being the same planting mix. One had been designed to be low and seen over, yet resistant to visitor access by being thick enough. The other was tall and thinner, designed to be a screen from the main lawn that partially hid Family Farm buildings and certainly hid visitors from either side. The low hedge alongside the path suffered an awful lot in one section from visitors plucking pieces off for the goats. It was very tempting to cover the whole hedge with mesh to stop it, but fortunately, the habit died out or the hedge became thick enough not to be so easily plucked. As it became stronger, it grew too much in spring, always a busy time in any garden and soon the meadow almost disappeared. With too much height, thick stems, and nasty thorns, the following winter the tree surgeons reduced the height again, shredding the thorny arisings into the nearby woodland. This was very handy and saved the horticulture team a lot of slow thorny work, plus the disposal of the arisings. The following spring, growth was even better and soon would need a light trim. No hedge cutting was allowed in nesting season, though. Strolling past one day, a memory came of reading about or maybe seeing a farmer thrashing a hedge with a long thin stick, breaking the new soft growth at its most delicate stage. It was not pretty to watch but was most effective. A new hedge cutting system developed using a sharp hand-pruning saw and a quick flick of the wrist. Soft growth was easily cut with no nests in that new growth (and if any seen deeper in the hedge easily avoided). It was so much quicker than with secateurs, and the rougher or more ragged appearance was more natural. It may have been another sneaky move, but well worth it. Only in the autumn when the leaves fell off could bird nests be seen easily – in a 75m length, there were maybe six nests, mainly chaffinch and blackbirds. It really showed the value of a dense hedge for wildlife and was perfect for the education team. There were noticeably more nests in the wider low hedge compared to the thinner tall hedge, despite far more passing traffic along the low hedge.

Some dog rose shoots soon sprouted a strange-looking growth – masses of green and red spikes in a 25mm ball. This is caused by a tiny little gall wasp *Diplolepis rosae*. Usually called robins pincushion (from its appearance, Robin is another name for Puck, the woodland or nature fairy) or bedeguar gall, the growth is caused by the feeding activity of several larvae inside the gall. Harmless to the plant, great to see, yet another little bit of nature that the education team could use. The gall wasps are almost always all female, needing no males to reproduce by parthenogenesis – another interesting point for the educators.

The meadow had deteriorated badly in a few short years, taken over by creeping buttercup and red campion. The other plants were still there but totally swamped. It was a meadow of yellow and red flowers only. So, that autumn, very gentle weedkilling started – very carefully and thoroughly editing out the unwanted plants and the large batches of the thugs. After that, a deep mulch of tree chips was laid, leaving a meandering path through the whole length for gardening access to allow hedging, weeding, or deadheading. The meadow had become a herbaceous border, albeit with only native Irish plants. Each plant was tended for weeding and deadheading as required. This worked exceedingly well. The buttercup disappeared, the red campion had been reduced to a few small clumps that were deadheaded rigorously immediately after flowering, so fewer seeds were produced. The other meadow plants now grew far better, especially the meadow cranesbill – from which seed was collected one year and a 1kg package was sent back to the supplier and bartered for other seed. This is the crux of any meadow – it needs to be managed, or it is not a meadow for long. A traditional meadow takes off a crop of hay and is then grazed. Fertility is removed. This manages the over-vigorous grasses and allows other meadow plants to thrive.

An herbaceous border of native plants, carefully edited of thugs and heavily mulched to reduce seed germination, this will be a constant work in progress, varying a little every year.

The fruit and vegetable garden developed into a useful area for educational classes.

Strolling along the path most days (especially in the early morning before visitors arrive and in winter), there are always birds present, getting seeds or insects from the meadow or hedge. Proof that it does not take much to increase the native wildlife in the garden if given conditions they need. Wonderful.

After several years, the rear hedge completely screened views of the farm from the lawn area, while the front lower hedge allowed some viewing into the meadow.

CHAPTER 10

Gorilla Rainforest

A Massive Task for The Horticulture Team

Inspiration always has to come first, and for the gorillas, it came from a very specific area – the Mbeli-Bai swamp in the Nouabale-Ndoki National Park. This is an area of undisturbed tropical rainforest in the Republic of Congo. There has been no logging here for a long time. Mbeli-Bai is a large opening in this rainforest, where many animals come to drink and feed. A study area for many years, especially for western lowland gorillas *Gorilla gorilla gorilla* and forest elephants *Loxodonta cyclotis*, the smallest species being only 2.4m tall at the shoulder. The concept for the gorilla zoo habitat was an island. It would have a deep moat around the edge, giving a very natural soft barrier between the visitors and the gorillas, without too-close viewing everywhere. This gave the added advantage of allowing the gorillas more opportunity to stay back from visitors if they wished. A series of 4m high ridges stretching out from the main covered viewing area, like outspread fingers on a hand, gave shelter from wind from any direction. It gave excellent screening if any individual animal wanted to be alone for a while and tremendous planting potential for the horticultural team. Wet areas would be different again between the ridges, and around the edge of the island, there was great scope for water margin plants. The visitor trail would meander around the island, with walkways partially over the water in places, crossing two streams with waterfalls linking back to the existing zoo lake. With lots of tree planting, visitors would eventually be standing in a woodland, looking out at the island to see the gorillas.

As mentioned above, the concept of a new gorilla habitat was broached the day after the Kaziranga Forest Trail was opened in June 2007. Creating these habitats depends on the master plan, and the timing of each depends on moving other animals around. With the gorilla habitat shortlisted as needing urgent improvement, where would they fit? It has to be in the African-themed area, of course, but there was no room. So, build the savanna habitat and make room for the rhinos within that habitat. This would free up the current rhino area, and that could be used then for the gorilla habitat. Sounds simple, doesn't it? Musical chairs with animals.

net was a very ready supply of information, though the website was still very new and developing. Other animals cause plant damage that non-zoo gardeners have issues with, such as rabbits and deer. These can plague country gardens, hard to fence out and have diverse taste in garden plants. So, check through the many lists of rabbit-proof and deer-proof plants. Immediately, it is obvious that they are not eaten for a very good reason. The vast majority are poisonous, often acutely, hence best avoided. If you make a rabbit sick because it has nibbled your *Kniphofia*, no one will complain. The gorillas are a lot more precious (sorry rabbits), and no risk could be taken. The other source of knowledge was casual observation – geese, for example, with a reputation of eating everything within the zoo, never grazed *Helenium*. In any farmer's field, you can often see patches of *Iris pseudoacorus* and *Caltha palustris*, both of which we had used on island edges to great effect around the zoo already.

One aspect of many poisonous plants is that the toxins involved are very bitter, and that makes sense. The plants are poisonous because they do not want to be eaten. There is remarkable chemical warfare constantly going on between insects, animals, and many plants. The insects and animals want food; the plants want to deny them. This was described above with the *Acacia* trees and giraffes in Africa. Often knowing one plant species is really poisonous can point towards another species in the genus that is not poisonous but is still too bitter to be eaten. One truism is that poisonous plants are only poisonous if actually eaten. If animals are put off eating them, they will not be harmed, and the plant will probably be left alone in the future. The real risk and it was a very real risk, was that some truly poisonous plants would appear. Tonnes of topsoil had been sourced from outside the zoo. Seeds could come in and not be noticed. A gorilla became sick in another zoo when one very dry summer caused the water level in the surrounding moat to drop enough to allow the gorillas to sample plants they usually could not reach as those plants were growing in the water. The plants included just one clump (no doubt self-sown) of water dropwort *Oenanthe crocata*, a particularly poisonous member of the cow parsley family. Its leaves look very like celery, a regular part of most zoo gorillas' diet.

A long list of trees and shrubs, herbaceous plants and water plants was drawn up. Each was checked in various poisonous plants books and databases on the web. Large numbers would be needed, so part of that listing exercise was availability. There would need to be variation in plant size too; from small trees to tree whips and shrubs, herbaceous plants in small pots or liners, and seed for the grass and meadow areas. A simple spreadsheet was developed listing possible plants, whereabouts in the habitat they would be planted, and which nurseries would supply them. Estimating the number required was very rough and based purely on how large the area was for each mix. There were four planting mixes – trees and shrubs on top of the mounds, herbaceous on the slopes, meadow and grass on the flat areas, and water margin plants along the moat and between the ridges. Plus, a lot of 'extras' to allow that natural look of odd plants here and there.

On the visitor side, the African Rainforest planting mix came into play. This had already developed along the main pathways past the savanna – handy to be ahead for a change. There was a lot more variety here as it was much more than a screen. Some of the plants used in the gorilla habitat were included on the visitor side and vice versa to make the vegetation seem alike, increasing the immersion effect.

It took a year of researching what plants to use and organising suppliers. At meetings with the builders, the wee point was gently raised about the time allotted to do the planting. Only the last two weeks in an eight-month schedule of work had been allocated for planting. Now, really lads, that is not going to work, is it? Much more time was needed for actual planting and to get the plants established. In fact, the more time, the better.

There was one other problem. The rhinos had a newly fenced hardstand within view of the visitors. The plan with that fence line was on one piece of paper. The line of the gorilla moat was on another. The two were never plotted on the same plan… So, despite regular pleas to look at both together, that is why the moat-side planted stand-off area alongside the rhino hardstand is only 300mm wide.

Builders began work in late 2010. There was a very large amount of earth moved, especially to make the moats, which had to have loads of reinforced concrete for the vertical side as it was also a weight-retaining wall if there was no water present. The land-shaping in the habitat had to be done in stages, with the builders slowly working their way out. The ridges were made one at a time, starting nearest the rhino hardstand. Once enough subsoil was in place, a 300mm depth of topsoil was brought in and spread. This was all placed incredibly accurately by the skilful heavy machinery drivers – a simple swing of the excavator bucket, tilt a little more, and the soil was spread evenly enough to plant up. The planting of trees and shrubs went ahead immediately, with the deliberate scattering of a few of everything a little way down the slopes to make it look more natural, a drift here and there. Then the machinery drivers even brought mulch over in a bucket that held about a tonne at a time. They ever so gently tipped it and spread it as they went, over the planting. All that was needed from the horticulture team was a careful eye on anything getting buried and a little spreading with a fork. Thanks to those drivers, the work went along much faster and with much less effort. Keeping the builders and the drivers as friends always paid off.

The water moat, with a vertical visitor-side concrete wall and a gentle slope on the gorilla side. Note the depression at the top, for holding a little soil for water side planting, ready for the waterproof lining. The concrete wall has a sloping top, much easier to make it disappear into the soil and planting.

With water in, the same area just after planting the gorilla side. The mounds have already greened up, being planted maybe a month or two earlier.

The mulch used was the 'zoo standard' at the time, but this project needed so much we had stockpiled probably 250 tonnes in an off-show area nearby, along part of the boundary with our neighbour, the President of Ireland. This was all done very discreetly, and no one noticed. At least, no one noticed that we heard. Having a well-rotted heap of mulch made from half manure and half wood chips, which was light and easy to spread and handle, paid off even more. Stockpiling that amount earlier was immensely valuable, for both the time it saved and for the benefits of mulching.

The slopes were planted next. A mix of herbaceous meadow and grass seed was sown as soon as topsoil was ready, then the other planting was done. But which grass to sow? There were so many different grass seed mixes around. Advice from UK zoo colleagues pointed out the benefits of RTF – Rhizomatous Tall Fescue. This is a deep-rooting, very tough-leaved selection, originally from southern France and northern Spain. Drought tolerant, good dark green colour, tougher leaf, and the best part was it had a creeping rhizome that filled in any gaps. A self-repairing, tough, drought-resistant grass cover. Perfect. The idea was that it would root deep enough not to be pulled out if the gorillas were grazing and the leaves were tougher than usual to resist damage. But – and there is always a but – it needed soil temperatures above 15°C to germinate well, which was generally after mid-May in Dublin. It would establish better if given more time than we could allow before the gorillas had access. So, a mix of RTF and annual ryegrass was used, which would give a quick cover that summer but would eventually become a tough perennial sward that would take a lot of abuse.

The valley bottoms between the ridges were lined with waterproof membrane, as used in the moat. Planting here would be at a distance from visitors and just needed to be unwelcoming to the gorillas – a wet soil without interesting edible plants, if possible. It was all prepared by the builders. The liner had to be well buried, or the gorillas would probably pull it up where they could. Levels were not easy to keep, nor was it worried about all that much. This resulted in most of the valley bottom not being as wet as initially desired, probably not wet enough to dissuade the gorillas from using them. However, there were some very wet patches at the lowest points. Irrigation had been laid in place to keep it all wet even if there was no rain, but it had to be used a lot more than anticipated. Lush growth would probably eventually make the valley bottoms drier anyway.

The gorillas were expected to roam around a lot, but the large flat areas were designed to be better for visitor viewing and to encourage the gorillas to be comfortable there. Regular scatter feeding during the day by the animal teams (usually as part of talks to the visitors) served to keep the gorillas nearer the best viewing areas. Equally, animal teams hiding some food early each morning would also keep them nearby, even though the gorillas searching through the vegetation would probably cause a little damage. Plants here would need to be very tolerant of grazing and gorilla foot traffic and, hopefully, be edible too. A typical hay meadow wildflower mix was used on any grass area – without any poisonous plants being included. Deliberate sowing of clover *Trifolium repens* was used for some of the main flat areas. It was more drought-resistant and very tolerant of abuse, as it turned out. This was very welcome.

The moat margins on the visitor side were vertical concrete, and no water plants would be needed there. Across the water on the gorilla habitat side, there was a narrow marshy edge along the water's edge of varying width with about 75mm depth of water (to discourage investigation by the gorillas who prefer to keep their feet dry), then a gentle slope covered in a strong mesh to enable gorillas to climb out easily if needs be. Hiding the mesh and softening the land-water boundary was going to be interesting. Which plants would do best? The land-water boundary is always a very dynamic habitat and has often been very challenging around the zoo and rewarding – if successful.

The biggest headache for this project was logistics, especially plant storage between delivery and planting. There was nowhere it could be delivered to and stored all in one go. Fortunately, there was no need or desire to as the planting was spread over many months and sourced from several different nurseries. Once we had the builders on board with our desired planting schedule, we could order the right plants in the right quantities for whichever area was being planted next. That initial spreadsheet listing of the plants was very useful again, as it was sortable by planting area. The whole habitat, both gorilla and visitor sides, was divided up into forty individual plots. One problem was that some trees were coming in during the winter as whips, young trees under a metre high, cheap and bare root. An option would be keeping the whips moist in a fridge or cold room until needed, which works well, but there was no guarantee of when they would actually be planted exactly. They either needed planting during winter or potting up. If potted, the roots would barely have got going before they were due to be planted. Tricky. Eventually, it was decided to have them delivered early in winter, pot them on, two or three whips to a 5lt pot, but regard it more as heeling-in in a pot rather than in soil or grit as usual. This meant at least they could be stored and moved without disturbance; only the actual act of planting would do that. Planting could proceed as normal while always trying to keep as much soil on the new roots as possible and water in very thoroughly – even in gentle rain, which raised a few eyebrows. The vast majority of the tree whips survived this. We used a small yard next to the bongo habitat for all other plants (water margins, shrubs, herbaceous liners). Laying plastic sheets down to hold water for the water plants, with a timber edge under the sheet to make a shallow 'pond.' A domestic water tank with a ball valve sitting in the 'pond,' kept the water level correct. Sprinklers on battery-controlled timer valves kept everything else wet. It worked too but was chaotic at times. Deliveries had to be booked ahead, the yard might be full, but more orders went out. If the planting had been delayed at any point, there would have been nowhere to keep the plants. Every two weeks, another delivery lorry or two would come in, especially while we were planting the visitor side. Those were generally larger plants and hey presto, we had a full yard again. This was all done before too many visitors came in. Unpacking and sorting ten different bamboos was always interesting. This became even more interesting when a student, who was intent on improving his plant knowledge, removed the one and only label per batch...

With very limited staff numbers, any labour-saving trick was often very useful. When root-balled trees came in for the visitor side, using a forklift to handle them, they were heeled in on several tractor-trailers by covering the root-balls with chippings from the tree surgeons and adding lashings of water. DTB. It was simple then to drive the trailer to the planting spot when it was needed and offload it with a mini digger. The hard part was loading them backwards in the right order, so first needed was last loaded. Plus, the order of planting was dictated by the builder's progress.

During the build itself, a few changes were needed. After much arguing, a drain was installed from the lowest point of the gorilla habitat to reduce or, better still, prevent waterlogging around the roots of a large oak *Quercus robur*. The drain had to go under the moat, so it had to go in before the moat itself. The visitor trail has a boardwalk over shallow water at one point, and the vertical concrete moat side was visible there. The horticulture team was not popular after insisting that the top 300mm of the concrete side of the moat had to be cut off. It did look far better on completion when water filled the moat, giving a continuous water surface.

Most of the logs that give the habitat so much structure came mainly from within the zoo. These were old or failing trees that were identified as needing removal over the past few years for safety reasons. They were not removed immediately but as and when needed. After all, it is easier to store them while still vertical... The large oak directly in front of the viewing shelter took four hours

A sick beech listed for felling. The heaviest branches were kept for the gorilla house.

to travel on a lorry within the zoo. It was moved from near the bongos, with a tree surgeon perching on it to get under any low tree branches along the way. The builders' crane could not lift it in place on its own – a large digger had to lift one end. The weight was only about ten tonnes, but it was to be placed too far away from the crane – the further away it goes, the lower the weight limit for the crane. All other logs were put in as the build progressed, and once in place, they made it impossible to drive any vehicle or even push a wheelbarrow from one end to another without ducking under or going around. But they looked good and gave an extra dimension to the habitat for the gorillas which was the whole point, of course.

They always look smaller when in place…

...do not underestimate the time and patience needed to get a large log from one place to another within the zoo...

The day finally arrived to allow the gorillas access to their new habitat. This was planned a little ahead of the official opening – just to make sure they actually went out and explored. This was fortunate, as they showed very careful attention to moving past their door at first. Most interesting was the fact that the girls went out first. The big silverback male sat in the doorway, calling the girls back whenever they found any regular food (hidden earlier by the animal team) and gratefully taking it off them. A little bit like the good wife nipping up to the shops for the husband watching tv... Watching the gorillas pushing into the vegetation was an education in itself – for the gorillas and any observers. Remember, the gorillas had been used to a simple rough grass area plus a few turf weeds, and now they had this vast array of new plants. Will it hurt if they pushed through the plants? Are the plants edible? For the first few days, the gorillas even avoided cardoon *Cynara cardunculus*, jumping over it if it was in their path. On the very first day, one female gorilla targeted the purple willow *Salix purpurea*. Not surprising as they would have been given several different species of cut willow as browse for many years. But not *S. purpurea*. She pulled the plant up first. It was from cuttings we had raised in the nursery and nurtured. Groans came from the horticulture team. Laughs came from the animal team. She stripped the leaves off and pushed them into her mouth, instantly spitting them back out and throwing the plant stem down in disgust. Cheers from the horticulture team. Moans from the animal team. Then came the tentative question, 'Is it poisonous?'

Within a few days, the gorillas were happily wandering around trying out new plants or more often, looking for food hidden by the animal teams. The wildflower meadow seed mix had certainly made the area colourful, almost too much, but as the grass established, the wildflower numbers reduced.

Over ten years, there had been a slow reduction in plant cover, especially among the trees and shrubs, partly from being eaten, often from being played with, breakages, and pulling up a little. There is a vast difference in damage depending on how often the gorillas use an area – the more use, the more damage. Comfort or ease of walking also come into this; a narrow ridge that was harder to walk on has grown far better. The differences add to the complexity of the habitat for the gorillas, making it far more interesting. After a few years, a male group of red-capped mangabey *Cercocebus torquatus* was introduced, sharing the habitat. This gave excellent enrichment for the gorillas but very significantly, it increased the browsing and damage to the plants, especially the large oak, fully mature and about 100 years old. The gorillas had largely left it alone, while the mangabeys immediately climbed it and started checking how edible the leaves were. Now oak contains a lot of tannin, which makes them bitter. New leaves are red as they contain even more tannin, to reduce insect damage. The mangabeys did not appreciate, understand or remember this. They kept trying that leaf to see if it tasted better than the previous few dozen... The gorillas followed the mangabeys, and it looked great to see them all high in the tree. Unfortunately, there has been significant damage to the poor tree from gorillas and mangabeys over time, losing much leaf cover in summer and so much bark on many of the branches that a lot have died completely.

There was so much learned from the plant selection, with great enrichment from some and a few pleasant surprises.

Trees and Shrubs Used

Caucasian wingnut *Pterocarya fraxinifolia,* potentially a very large tree, is often seen planted on a lake edge. A member of the walnut family Juglandaceae, which are well known for various toxins, but *Pterocarya* was not listed as poisonous. The large pinnate leaves are noticeable even from a distance. Once mature, there are long pendulous racemes of small green flowers, followed by green-winged fruits, decorative too. Trials with suckers pruned off and given to goats (a real test) proved that the leaves were not readily eaten and any thrown into the gorillas' previous area as a test were always left well alone. Three root-balled trees of 3m tall were planted and initially given hot wire protection to stop any climbing. Some of these were very carefully positioned to help screen the view of the house from the decking. Fifty whips were planted, cheap and easy and without protection. The protected three root-balled trees grew well enough, with some damage to branches but with competition from grass and herbaceous plants as ground cover, they are still slow. Equally, the whips have grown slowly with competition, and not surprisingly, several have disappeared. Any that do survive will hopefully start suckering. This is another good reason to use this species, as it can form very dense thickets in time. The hope is that if the suckers are not liked, the tree may not be climbed as much and the whole area avoided. Time will tell.

Purple willow *Salix purpurea* was immediately targeted as it was recognised as willow, which is normally edible. *S. purpurea* was selected for planting as it is not eaten by rabbits, due to the higher salicylic acid content making it very bitter.

The salicylic acid in willow is a well-known plant source of aspirin. Years ago, the doctrine of signatures used for medicinal guidance said that trembling willow leaves were good for a trembling fever, and they were. No doubt, each gorilla and mangabey had to discover that this species is not nice to eat, and some plants were lost. But out of the 100 or so planted, enough have grown to 3m bushes to give a good shrubby look to some areas. After a few years, once large enough, *S. purpurea* started flowering, giving masses of catkins along the thin twigs, typical of a willow. Both the gorillas and the mangabey were seen one spring in the middle of the bushes, pulling the shoots down, stripping off growth, and letting go. Only with binoculars could anyone see what they were doing. Very carefully, the catkins – only the catkins, were being removed, no young leaves or bark and immediately popped into waiting mouths. Individual technique varied, maybe one catkin at a time, maybe a few. The gorillas were even seen with green chins from the pollen. Obviously proving a tasty and totally unexpected mouthful. The question remains, why? Is the catkin a temporary growth that the willow does not bother forming the bitter salicylic acid within? Is the nectar sweet enough to hide the bitter flavour? Most likely to keep the nectar sweet for pollinators, the bitter salicylic acid is not produced in the catkin. Once flowering stops, the willow is left alone again, until next year, apart from incidental damage from playing and climbing.

The gorilla seemed to be waiting for the willow catkins to get big enough, while his neighbours attacked the *Gunnera* buds. Who taught who?

One challenge encountered watching this interaction was the reaction from one of the gorillas. Whenever some of the horticulture team came around with a camera and binoculars to see what was being eaten, the gorilla immediately stopped doing whatever it was engaged in, looked up, watched, and shadowed the photographer all around the moat, always looking directly at the photographer. But only when the photographer was wearing the zoo uniform or was alone. If they went around with a group of visitors, they were not noticed so easily. The best pictures were taken while taking guided tours around. Indeed, the observer does sometimes change the observed.

Various cultivars of firethorn *Pyracantha* were planted hoping to reduce the gorillas' use of certain areas, such as around the base of the large oak, which was not initially intended for climbing. Despite the vicious thorns, not all survived; some were browsed and young growth was eaten – always a risk as the thorns are soft at first. It was all good enrichment, even at the loss of some plants. But some did survive, and these then gave further enrichment as the berries became sought after. The gorillas would carefully tease a couple of berries out of a thorny bush, taking time to avoid the thorns, going back for more, one at a time. Never very many to harvest, but tasty enough to be worthwhile.

Box honeysuckle *Lonicera ligustrina* var. *yunnanensis* (*nitida*) is an awful thug of a plant in a garden, making a dirty hedge that needs constant clipping. It is easy to grow, cheap and evergreen. A few pots went in, and sure enough, their thuggishness overcame the gorillas. A large clump is now a dense screen and windbreak. It has become a favourite spot for hiding under and facing south, which makes it a suntrap. If not pruned, it will flower, though the flowers are small and insignificant. These are followed by small black berries, which the gorillas found worth hunting out too, but as with the *Pyracantha*, the berries were picked one at a time, with very little damage to the plant. Ignore the common name for this plant. It may be in the honeysuckle genus, but the similarity stops there. Better common names include poor mans' box. The perfect description as it grows so easily and is so cheap, but box *Buxus sempervirens* is a far nicer plant, a shame to tarnish it with any connection to this thug.

The common hedge plant – in milder parts of Ireland anyway – New Zealand broadleaf *Griselinia littoralis* has been a great survivor. The thick, glossy, leathery leaves are not palatable. Again evergreen, so it is a valuable screen and windbreak. It can become a large tree, though maybe not in such an exposed spot. But, nothing ventured... Many years previous, this had been planted on lake edges with lots of easy goose access, and even though planted as small 'whips' and suffering a little bark damage in winter from the geese, the leaves were left alone completely. The bark was stripped in places. It is hard to imagine geese doing it, but by turning their heads and running their beak up and down, they managed to do so when grass was very thin in winter.

The even more common and equally horrible as a hedge, privet *Ligustrum vulgare* is left alone, apart from an occasional nibble of the berries. All parts of the plant (especially the berries) have potentially toxic glucosides, but it has been used in many zoos with no problems. The plant is almost untouched, with only an occasional berry being carefully picked.

Gorse *Ulex europaeus* is an awful weed of a shrub, flinging its seeds around to form dense impenetrable colonies of tough, very sharp needle-tipped leaves. Animal team members had seen various primates eating the flowers in other zoos and had specifically requested some to be planted. Very reluctantly, a few plants went in, but only the double form *U. e.* 'Flore Pleno,' so there were no seeds. Praise be. Why reluctance? Gorse is such a common plant on rough or high land that it just looks out of place – the gorilla habitat (and the chimps nearby) looked very much like the Dublin Mountains not too far away. As expected, it has proved gorilla and mangabey-proof, but it will slowly get bigger and simply not look right, even if it does give a little enrichment as the flowers are gingerly picked off.

Strongly scented plants had proved resistant to browsing in other parts of the zoo. Peafowl, the bane of the gardeners' life for years, had completely left the Mexican orange blossom *Choisya ternata* alone. The leaves are very strongly scented, and it is a tough evergreen. Here the plants are left alone, but it would be interesting if the flowers were taken, though that has not been observed yet. *Choisya* is in an interesting family, Rutaceae, and includes all the citrus group, oranges and lemons, etc. Hence the common name as the flowers are very similar in shape and scent. It also includes rue *Ruta graveolens*, a herb not often seen and treated with great care as the leaves can cause an allergic reaction, skin blisters, or gastric issues in some people – one not to plant for sure. More interesting still, cats do not like it and tend to avoid it if planted in gardens.

Equally scented and more pleasant to our noses is the laurel or bay tree *Laurus nobilis*, used in cooking. A tough, leathery leaf may not be the most palatable with a strong scent. It has sesquiterpene toxins which in other plants may cause allergic reactions in higher concentrations or different formats. It is very slow to grow with competition from surrounding grass, etc. Some have survived the first few years. In the wild, it would reach 15m tall. It would have been very dominant around the Mediterranean over 10,000 years ago, slowly reducing as the climate changed and becoming drier. Symbolically used as a sign of status or victory in ancient Greece and Rome, it is still evident in the terms 'poet laureate' and 'resting on your laurels.' The common name, laurel, has been used for many other plants worldwide, most confusingly.

A twig from *Laurus nobilis*, with the bark stripped off and the top leaves still intact. Normally when given browse material, the leaves are eaten first rather than the bark. Great enrichment either way.

Prickly plants used included various *Berberis* spp, *candidula*, *darwinii*, and *julianae*. Some have survived in more awkward to access spots. Unfortunately, *Berberis* are normally sold as small plants, and the majority seemed to have been swamped or damaged. The gorillas were keen to try the new growth and often pulled the whole plant up as they tried to 'pluck' the growth off. With enough years' notice, some larger plants could have been grown on, but there were more

leaves to peer underneath, often picking out seedlings of dandelion *Taraxacum officinale*, which had been noted as a favourite plant by other zoos. The success of the *Cynara* here and the appearance suggested its use on the orangutan habitat (see below).

Chilean rhubarb *Gunnera manicata* (but absolutely no relation to real rhubarb *Rheum*) had been used in gorilla habitats in other zoos and largely left alone. It is such a magnificent plant, used much more in other parts of the zoo, and gives such a tremendous show that it had to go in. The wider parts of the watery margins of the moat were perfect for it and the shallow water would give some protection from the gorillas too. Some damage was noted, but a regular spring sequence for a few years was to see both mangabeys and gorillas seeking out the flowering shoots before they grew much. Once growth starts to swell the large buds, both animals can be seen gently teasing apart the stiff bracts that protect the new shoots, until they find the flower bud, which is normally at the top of the growth anyway. This is then eaten, delicately picked apart while still on the plant, or pulled off completely as a vegetarian take-away. The leaves are left alone, apart from damage from playing, but growth is less magnificent without a gardeners' lavish application of deep mulch each winter. *G. chilensis*, the awful invasive species found in several parts of the west coast of Ireland and elsewhere, has been eaten by native people in Chile. This is eaten by stripping the prickly skin off the stalks, dipping the ends in salt and eating like celery or made into marmalade – a new cottage industry maybe awaits in Ireland?

Pendulous sedge *Carex pendula*, a very common native, vigorous, self-seeding, empire-building thug that often causes plant-aware visitors amazement that it is actually 'planted' – a real survivor in often very tough conditions. Flourishing with wet feet, a perfect plant for the moat edges, with a good dense root mass to keep it in place. The gorillas discovered that the flowering shoots pull out of the plant if tugged hard enough, before the flowers even open. The basal 100mm of the stem is then eaten or at least chewed, presumably the best or sweetest part – the rest is discarded. The mangabeys soon followed suit. Which animal really found it first is guesswork, as with many of these observations.

As part of the grass seed mix, red clover *Trifolium pratense* was used, generally throughout, but especially near one of the feeding areas for visitor talks. The intention was to have a drought-resistant element that would hopefully stay green. *Trifolium* is a well-known fodder plant, a standard component of old-style hayfields, and is regarded as good for general health by herbalists. It was no real surprise to see it eaten. However, the animal team reacted with alarm as the gorillas suddenly found this new edible plant. In late afternoon as they were called in, they would gather armfuls. The animal staff was concerned it may be harmful. The query was one that only an animal team member could voice. When asked on the two-way radio for a description of the plant, all that could be offered was that the plant had green leaves... There is a downside to being popular, though. Each plant was sought out and carefully harvested, while plain grass next to it was ignored. The poor clover was slowly reduced. It is harder to find but still exists throughout the grass. There are little bits here and there, giving great cause for much walking, looking, and thinking. Excellent enrichment for the gorillas again.

Red clover actively sought out once recognised and thoroughly enjoyed.

Other zoos had reported gorillas seeking out dandelions *Taraxacum officinale*. This is well known and eaten by humans too, in salads or made into wine from the flowers. Though it should be taken with care as it is a diuretic, hence the old common name pee-a-bed. Soon enough, the gorillas were seen looking, peering under bushes, even gently lifting the leaves of the *Cynara* to see underneath. Though binoculars would be needed to see what plants they were after. Small seedlings were even plucked and eaten, weeding them out – gorilla gardening.

Slow foraging was great enrichment.

Vigorous comfrey *Symphytum officinale* was eaten eagerly, despite the rough texture of the leaves when fresh. This has been used as animal feed – more so with organic farmers – and many animals eat it readily, but to make it more palatable, the leaves can be left to wilt a while, making the stiff hairs less so. Years ago, when offered to a very young Siamang gibbon *Symphalangus syndactylus*, it had thrown the comfrey down with a look of absolute distaste after a small nibble. The mother gibbon had picked a leaf up though and chewed it thoughtfully with the youngster watching. After a few minutes, the young one picked a leaf up again and chewed it, copying its mum, but the look on its face was not one of happiness. Recognised as good food, the gorillas enjoyed it but again, too much so. There was very little growth after a year or two. Though it is a very strong grower, daily cropping would have reduced it, especially when newly planted and in competition with grass.

Geese had been a useful guide to plant edibility over many years – they eat almost anything, remember. After much planting and lake-edge fencing over the years, the geese were limited to a much smaller grass area, within which seedlings of elecampane *Inula helenium* appeared and were left alone. Interesting. Like *Laurus nobilis* and *Cynara* above, the genus *Inula* has similar potential allergens, thus preventing browsing. Cue using this in the gorillas too. Even better is *Inula magnifica*, as its name suggests, a magnificent plant with larger leaves and growing to 1.5m. 100mm yellow flowers are eye-catching. The same plant used on the visitor trail looks good just as leaves and great when flowering. With intense competition among the meadow mix in the gorilla habitat, they are not as vigorous, but still survive.

Edible, palatable or not nice...much time spent checking, the main purpose of the gorilla side planting, enrichment.

Water Margins Along the Moat

The water edges were expected to be a challenge, and they were. From planting in mud while trying not to slip into the moat, to getting enough soil around the roots to anchor them in enough, they were challenging. A wide range of species was tried; some did very well, some not so well. Almost the entire edge was covered, which hid the plastic mesh under the water and looked far better too.

The meeting of water and land is always the most dynamic aspect of habitats, changing rapidly as plants succeed or fail. And there had been no chance to trial any of these with the gorillas to see if they would be eaten. However, we did know of tough animal-resistant plants from the lower lake planting.

Marsh marigold *Caltha palustris*, the native plant that brightens up many wet areas in spring with buttercup-style yellow flowers, and *Caltha polypetala*, a giant species – imagine *C. palustris* on steroids – were both suitable. There seems to be much confusion with *Caltha* naming; this giant was first acquired as *C. palustris tyermanii*, then seen in a botanic garden as *C. polypetala*. Checking plant lists, there seems to be a range of synonym; for instance, *C. palustris* var. *palustris* or *C. palustris polypetala*. The smaller native *C. palustris* struggled in competition, though the larger *C. polypetala* is a wonderful strong grower and looks good even from 5m away as it is so large. All parts are about double the size of the native species, with metre-long shoots and flowers 60mm across. Both species spread by rooting at each node, making this a very quick spreader and very handy. As with all the buttercup family, they contain the glucoside ranunculin. This could cause allergic skin reactions and gastric issues, but animals realise this. Think of all the fields that have marsh marigolds or buttercups happily growing with grazing cattle, sheep, or horses and all without animal health problems.

Yellow flag iris *Iris pseudoacorus* was another well-tried and trusted plant for water edges, used on many primate islands in the lower lake. A real thug, the rhizomes will creep over each other, forming a very solid mat that resists erosion superbly, and nothing eats it. Wonderful. Attention was needed when placing it, as when fully grown, it would be tall enough to hide the gorillas, though it does die down in winter. Odd spots along the edge would be fine, but a long line could be a problem. In time, it will spread, and some editing may be needed.

Iris pseudoacorus in great form, but too many and the views onto the island would be compromised.

Other water-loving Iris were largely left alone, *I. versicolor* and *I. laevigata*, as both give splashes of colour in early summer. The gorillas have been seen breaking off and checking the seed pods but throw them down soon enough when they find out they are not tasty.

Common reed *Phragmites australis* did well, but care was needed in its placement, as with the *Iris* above. The name again is a giveaway – *phragmites* from the Greek for hedge. It is not evergreen but persists as it has dead stems all winter and is tall enough to hide the gorillas. They were only planted where views would never be wanted. With the perfectly still water in the moat, you could really appreciate the *Phragmites*' erosion control and soil-building possibilities. Once well established, the plants send out runners – long black roots which float and root in wherever they find new soil. Together they would cross and re-cross any muddy areas, forming a web of rhizomes that trap soil particles very effectively. Seeing these black 'ropes' heading across the open water of the moat (about 5m) did cause a little worry. This was easily resolved by cutting them off at the gorilla side – though not with a spade as one animal team member did once... Remember, the whole moat is waterproofed with a heavy liner that a sharp spade could damage...

Water plantain *Alisma plantago-aquatica* is a really good-looking native but not showy. It has large paddle-shaped leaves just above water level and a spray of white flowers to about a metre tall and only 15mm across, but a good clean white. Not much to see around the gorillas as too far from visitors, but the floating seeds have dispersed around the moat, over the waterfall, into the upper lake and slowly down to the overflow... Thus, to the lower lake, where one grew in a very dry year on mud as the lake level dropped. It flowered and seeded like mad again a little further downstream. Great to have volunteer plants that look good.

Other typical waterside plants are the two species of bulrush *Typha angustifolia* (narrow leaves) and *T. latifolia* (broad leaves). They are both native, aggressive spreaders, and to some extent, both keep their dead leaves and stems through the winter. Again, not too many were planted where they might block the view. A minor worry was the possibility of the 1.5m stem falling into the water and potentially encouraging the gorillas out too far. Their weight would quickly make them sink, though the mangabeys may use them as a jump off point. No planting of these at any narrow points of the moat – just in case.

One of the best survivors is bogbean *Menyanthes trifoliata*, a native plant that loves shallow water, rooting into and helping to stabilise muddy soil as it goes. Low but vigorous growth causes no animal visibility issues. It was perfect for the moat edges. The thick stem creeps along, forming extensive mats that are sprinkled with small white stars that are a pleasure to see when in flower in early summer. Bogbean though? Alright, it grows in bogs, and the leaves do look a little like a broad bean – with some imagination. The leaves are very bitter – hence the lack of any noticeable nibbling, no doubt – and have been used to flavour schnapps. And that's another Irish craft industry possibility...

All of these plants and several more species were planted almost at the last moment. Most of the moat had a shallow planting area of varying width, which could hold a little water. The plan was that the gorillas would not go into the shallow water and thus be even less likely to go into the deeper moat. But the moat had to be full of water to get the plants really happy, and that was one of the last jobs – the builders did not want a water hazard right next to their work. Having been planted in midsummer, plant growth was generally good, but many plants had hardly gotten established by autumn.

Dublin Zoo is blessed with two lakes that encourage small flocks of native waterbirds to fly in. It is always interesting to see what arrives. Well, what arrived at the gorilla moat that first autumn and winter was a mob of coots *Fulica atra* and mallard ducks *Anas platyrhynchos*. They thoroughly enjoyed pottering through the mud of the moat edges, nibbling plants and insects alike – and often dislodging the barely established planting. With no way to prevent it, the damage went on, but fortunately, enough of the planting survived to grow the next spring. Praise be, we had used a wide range of plants. A few edges became much barer than desired, but the survivors will colonise gently.

Coots were very happy to nest nearby. Note the verdant bogbean *Menyanthes trifoliata*.

Not all was lost, though. Dislodged pieces of brooklime *Veronica beccabunga* (possibly from the Danish *bekkebunge*, brook bunch?) floated gracefully along the moat with the flow generated from the water circulation pumps. It is a vigorous plant with roots at each node and is a tough member of the plantain family, including some very successful weeds. It has nice blue flowers too. At the waterfalls, there were very random rocks that trapped many particles of debris, leaves, and twigs. That needed watching in case it made a dam and the water level rose, to escape somewhere else – perhaps through the planting and over the road, making a grand mess. But the *Veronica* got caught up, and with the daily flow keeping it wet, it took root. Eventually, it formed a thick mat hanging down over the rocks, with water flowing over, under and around it. It was the most natural-looking effect. When commented on, it always got the response, 'It took ages to get it right…'After a few years, the pumps broke down. It took ten days to get them fixed. No water in means no water out, and the waterfalls ceased to flow.

The brooklime was eventually supplanted by watercress *Nasturtium officinale* (syn *Rorippa nasturtium-aquatica*), which was equally as happy to float around until a new home was found. It is also more vigorous than the brooklime.

Waterfall? Maybe watercressfall is a better name…

One of the most pleasing aspects of all this waterside planting was the unplanned wildlife that appeared within a year. The zoo lakes had always had some native wildfowl, but with no geese damaging the water margin plants, there was much more growth that developed into a perfect watery habitat. Fish, sticklebacks *Gasterosteus aculeatus*, appeared very soon – the pumps that pulled water from the lake would have pumped through small fry – a tough journey, but some made it – and with an overflowing waterfall as the only outlet, they could avoid going out again. Various ducks also made the moat home, as did the coots and moorhens *Gallinula chloropus*. None of them did any real damage in small numbers once the plants were established, and they had safe nesting too. It gave great pleasure to visitors and staff to watch the young going around in season. The real gems were little grebes or dabchicks *Tachybaptus ruficollis* when breeding. In the wild, these are incredibly shy birds. The name *Tachybaptus* comes from ancient Greek, *takhus* meaning fast, and *bapto* meaning to sink.

Normally once seen, they dive and then disappear into vegetation. But around the gorillas, especially on the walkway partially over the water, they have become accustomed to being watched. If you stay still and the sunlight is at the right angle, you can see the parents dive down to catch small fish and bring them back for the young, who sometimes take fright anyway and climb on top of a parent and burrow under their wings to shelter. The adults may sometimes turn their heads sideways to look up when they surface, just to check! But to be within 3m of such usually nervous little darlings is truly magical.

Normally nervous dabchicks would happily feed even though they are within a few feet of people – once the people stayed still. It is magical to see the wild birds so close up.

Visitor-side Planting

Think back to our inspiration for this area – an opening in the rainforest. Therefore, the visitor path should be densely planted with much tree cover and provide views across the moat towards the gorillas where desired. Hefty clumps of vegetation would help screen visitors from each other as they look along the winding path. A careful balance would be required. Such dense planting would help to screen the visitors from the gorillas, making them feel more at ease. Remember the five freedoms – the gorillas may want to move away or be out of sight, and the planting means they can. Large-leafed lush vegetation was going to use a lot of herbaceous plants – all that lovely new growth would look good each year, but it would need to be cut down each winter, controlled, and managed. A balance would be needed. Not too much could be cut down, or there would be gaps galore, meaning excessive viewing. Evergreen planting would be essential in part for a natural-looking mix, the same as in any habitat. The trees used would need to be a good size eventually but not too large and plenty of them. In reality, they would need to be tolerant of pruning, even coppicing. The risk of branches getting too near the wrong side of the moat, or falling and bridging it, was an important point in the plan and it had to be considered from the start.

The largest number of trees used were tulip tree *Liriodendron tulipfera*, from the Greek *leirion* lily and *dendron* tree and so we have the lily tree bearing tulips. Some were ordered as root-balled 4m tall specimens – about the largest that the team could handle in tight planting spots, but with another 100 as whips – cheap, under 1m tall, easy to plant, and quick to grow. If some had to be cut down, this was less traumatic for the team with the idea that coppiced woody plants often grow quicker and with significantly larger leaves than usual. Indeed, some have done that nicely. There was a constant balancing act between growth and viewing. Often trees would need a gentle crown raising, removing side shoots to allow viewing past the narrow trunk. Explaining this to non-plant people always got the comment 'just cut it down,' which does not work as several shoots arise, and they would be more of a block than the one trunk. The leaves are uniquely shaped – three-lobed and foreshortened, as if a very artistic person had designed them. This has given a very distinct appearance to the habitat, which was the whole idea.

Gentle crown raising and removing low side shoots of *Liriodendron* and *Populus lasiocarpa*, enables viewing without reducing the tree canopy itself. White *Zantedeschia aethiopica*, purple *Iris laevigata* and long, flowering shoots of *Beschorneria yuccoides* make it better still.

Keeping the purple theme for some plants in the African Rainforest led us to the hybrid Indian bean *Catalpa x erubescens* 'Purpurea,' with large leaves that are a good deep purple when young. A superb bonus is the large flowers in summer, in good upright panicles, like horse chestnut *Aesculus*. Most flowers are well above visitors' heads. How many actually see them? Like the very similar *Paulownia*, *Catalpa* can be coppiced, resulting in larger leaves and vigorous upright growth. This cultivar is one of several hybrids between the North American *C. bignonioides* and the Chinese *C. ovata*. Ultimately, this can be a large tree of up to 20m, but it is unlikely to be allowed so near the moat. Coppicing will be interesting in a few years.

As the visitor path disappears around a corner, *Catalpa x erubescens* 'Purpurea' in full flower makes the trip a journey of discovery. Yellow *Inula magnifica* is always a good show. The large leaves of *Gunnera* almost swamp the purple elder *Sambucus nigra* 'Black Lace.'

Interesting trees are more expensive, so cheaper fillers of larger sizes were used with the horrific possibility, or most likely, the probability of needing to cut them down in a few years. Several oriental plane *Platanus orientalis* were used. Planted in the stand-off with a restricted root run, they will never get to their potential 25m, but with their glossy digitate leaves, they made an immediate impact, and they grow rapidly too. The oriental plane is native to 'the east' as its name implies, from the Balkans eastwards, but often planted over the millennia, so it is hard to say for sure. Great for shade in a hot sunny climate. A more delicate-looking leaf than the usual hybrid plane or London plane, *P. x acerifolia* is a cross between the eastern oriental plane and the western (from N. America) *P. occidentalis*.

Large-leaf trees were needed, but unusual ones. None more so than the Chinese necklace poplar *Populus lasiocarpa*. This lovely tree from temperate China has leaves that can reach 35cm long by 25cm wide. They are very different and noticeable. The leaf stalks and midrib are a rich red, which adds to the effect. The flowers are easily missed as they are catkins (poplars are in the willow family, Salicaceae), green, and up to 100mm long if you see them. But the seeds are produced in vast quantities of fluffy down that travel on the breeze and collect like snow on path edges in July. Hence the specific name, *lasiocarpa*, meaning 'woolly fruited.' This tree was first planted in the Kaziranga Trail but suffered from too much competition with the bamboos. Seen and admired in other gardens, it had to be tried again. A few large root-balled specimens and quite a few whips were planted. The whips rapidly overtook the root-balled trees in a few years, which often happens.

Elderberry is such an easy plant. It grows quickly and is tolerant of hard pruning. *Sambucus nigra* 'Black Lace' with its deep purple, daintily cut foliage and pinkish flowers, fitted in perfectly. Perhaps it is too vigorous but is easily pruned back every few years if felt to be getting too woody, which then results in rampant regrowth.

Under and between the trees, there was great scope for many herbaceous plants. Again, we were seeking options with large leaves or maybe purple leaves. *Phormium cookianum* 'Black Adder' gave us a solid purple visual barrier, fully blocking the view of the gorilla island as you enter the trail from the savanna. This is perfect, making people move along to get to a point where they can see anything other than plants, reinforcing the discovery element of walking through the zoo. It is not as vigorous as the normal *P. tenax*. Praise be. That can be a real thug if happy and not an easy plant to prune back. Two other differences are that *P. tenax* is hardier, though even that can suffer in a bad winter and the interesting flowers are not on such long stems – a good point as here they would lean over onto the visitor path or partway across the moat. From New Zealand, *Phormium* gives an important and strong material for cloth or rope making. The name comes from the Greek *phormion* meaning mat. Seed was first brought back from Captain Cooks' first voyage by Sir Joseph Banks.

We had used *Cynara cardunculus* in the gorilla habitat itself, so a few had to be planted in the stand-off area to keep the theme throughout, emphasising the immersion principle. The difference was that in the stand-off area, we could look after the *Cynara* properly by keeping weeds away, pruning, and mulching as required. What a great plant. Spring growth goes from soil level to 2m with large blue flowers like a thistle on steroids. Leave the flowers intact, all the leaves will die off and the stems remain bare all winter – a wonderful skeleton. Winter winds often blow them over, and they are tall enough to fall onto the visitor path and be a nuisance. Sneaky comes in here… Immediately after flowering, cut down any plants near the front to soil level. New leaves grow like mad and you have 1m tall silvery foliage through until next spring. Leave the ones at the back out of the way, and you have the architectural taller dead stem. The best of both worlds. All you need do is feed the brutes. Lovely.

The silvery mounds of large-leafed *Cynara* look good all winter, providing great early structure to the planting.

A wetland edge should have reeds, and what better than the giant reed *Arundo donax*. Another great foliage plant, growing to 3m or more each year from ground level and is the epitome of lush growth. Around the Mediterranean, you can see drainage ditches full of it. There are only green leaves and stems here, but it is a great statement in the landscape and a great view blocker, hiding more distant views until around a corner. Annual pruning to ground level stops flowering, but they look untidy and tattered if not cut down regularly, and they actually look lusher with all that new growth. *A. donax* has great potential as a biomass crop as it has been reliable for many years and enhances the soil. It loves water and has been used as part of water treatment processes. If all that doesn't make it worth growing, it is still the primary source of reeds for modern woodwind instruments. The stems make flutes and panpipes too.

Mentioned already and planted in the savanna is *Melianthus major*. It is a perfect plant here with its large leaf. It is unusual, not seen too often, and has a good flower of interesting form.

Beschorneria yuccoides is another most unusual-looking plant from Mexico. It has large leaves like a *Yucca* – hence the name – and is in the *Agave* family. The 2m tall flower spikes are produced readily when growing well, and they curve over as they grow, almost triffid-like in appearance. Really cold winters may affect it, especially if freezing snow lays on top of it, but it seems to be more frost tolerant than thought and regrows well.

With showy 70mm red/purple flowers, *Lobelia tupa* from Chile is a giant perennial relative of the hanging basket favourite. It dies back to ground each winter. Vigorous growth soon has the flowers at the top of 1.5m stem, with large leaves clothing the stems. These are long curved tubes, with the stamens held above the flower. The flowers are pollinated by hummingbirds with an equivalent curved beak, the feathers on top of the head would carry the pollen well. *Lobelia* is an incredibly diverse genus. In Africa, there are giants such as *Lobelia deckenii* on Mt Kilimanjaro, with a rosette of leaves that hold water which in freezing conditions protects the centre from getting any colder – the ice acts as an insulator.

Already discussed in the gorilla habitat, more *Gunnera manicata* had to be included along the visitor trail. The perfect spot was at one of the waterfalls with a 5m stand-off area, and there was enough room for the *Gunnera* to grow and not overflow onto the path too much. With a few plants on either side of the outflow, the stream disappears each summer, flowing under a high canopy of *Gunnera*, a lower canopy of wonderful royal fern *Osmunda regalis* and a carpet of the *Veronica* mentioned above. The plants had to fight it out to get where they are, but again, if asked, the comment must be 'it took ages to get it right.'

The purple theme continued with leopard plant *Ligularia* 'Britt Marie Crawford.' Its deep purple leaves and large yellow daisy-like flowers grow to about a metre tall. Dying back in winter leaves a gap proving the value in varying the planting to include a diverse mix, but the new dark maroon leaves in spring are worth it.

Some areas would have a lot of almost constant viewing over the planting, where only short-growing plants could be used. This could perhaps be broken up with trees that could be viewed under, by raising the crowns gently. Such pruning could be a very minor snip of half a side shoot – always trying to favour the tree a little, try to keep the side growth as it encourages main stem thickness. Large numbers of black mondo grass *Ophiopogon planiscapus* 'Kokuryu' gave a permanent cover all year round of yet more purple-leaved plants. A member of the *Asparagus* family and originally from Japan, this plant slowly creeps, forming a dense mat. It can be a pain to keep weed-free if any annual weeds get near it. Though it is too short to mulch thickly, fine particle bark should work. Flowers borne on stems of only 150mm long are noticeable but very small and not showy, especially at ground level. The berries that follow are far more so, a very dark violet-blue, almost black colour.

Giant inula *Inula magnifica*, planted in the gorilla habitat, is a massive presence in the stand-off area too. Dying down fully each winter, new growth in spring reaches 1.5m quickly, and the large yellow flowers are really impressive en masse – a solid block of vegetation as a screen, topped by colourful flowers. Perfect. It is easy to grow – deadhead after flowering to keep tidy and make the planting look greener, hard prune to ground level each winter, mulch to feed the brutes and keep weeds out. Simple.

Tropical appearance demands large leaves, so a few bananas had to be used. The Japanese banana *Musa basjoo* is readily available and grows well, but sometimes it needs frost protection in winter. This is the only way to have real height – it will get to 2m or more if happy. If heavy frost kills the stem above ground, it will regrow from the base, from side shoots or 'daughters.' Flowers appear when the stem is mature enough, and even small fruit appear, but that stem will then die down. It is herbaceous, not a woody plant. In Japan, this banana is grown for fibre – the fruit is inedible. Confusingly, Manila hemp is also from another banana species, and real hemp comes from cannabis plants chosen for fibre, not drug production. To really confuse people, tell them that botanically the banana is a cylindrical berry, while strawberries and raspberries are botanically aggregate fruits, not berries. Cucumbers are cylindrical berries too…

For evergreen tropical looks, there had to be some bamboos. Our decision to have large-leaf species for all Africa meant we were using potentially vigorous growing rhizomatous thugs in a narrow stand-off bed. Sure enough, within a few years, there was a need to reduce the amount and spread of the bamboo in places. Large-leaf bamboo *Indocalamus tesselatus* with, as you might guess, a large leaf to 50cm long and 15cm wide is great and not too tall in restricted growing conditions during the average Irish summer. Crushed up against the essential stand-off fences, it becomes less ornamental. The leaves swamp other plants meaning much digging out was needed. *Sasa kurilensis*, *S. palmata* 'Nebulosa,' and *S. tsuboiana*, are all good shorter species. They are tough enough to form a barrier in places, as long as no one really wants to get through them. They were used in odd spots and as fillers, often where visitors would be passing by rather than looking through or looking over. All have large enough leaves to look a bit tropical.

Japanese temple bamboo *Semiarundinaria fastuosa* is well known for its vertical growth. This is very useful when you want a dense evergreen screen but no 5m canes drooping over a path. It is perfect around the waterfalls. There, they make a visually solid screen where the path bends, completely hiding the views in front, keeping the idea of a visitor being on a journey of discovery. Imagine this in its native land – planted around temples, where it grows to 7m, straight up and solid, hence *fastuosa*, meaning proud.

The stand-off fence line – an essential part of the habitat – was a great place to grow climbers, but only wherever visitors would not be standing looking into the habitat. Otherwise, the plants would screen the view too much and the visitors would damage the plants pulling them out of the way. Chilean glory flower *Eccremocarpus scaber* was grown from seed which gave lots of small plants and no definite plan for where they should go. It was planted along parts of the stand-off fence near the waterfalls. This gave great cover to hide the fence and a marvellous colourful display too. The small red or yellow pendulous tubular flowers are produced in abundance but are only about 20mm long. They are followed by puffy seed pods. It is good to have a show of them as they are not planted often enough, despite being cheap and easy to grow from seed. Though not meant to be hardy, it came through several winters with little damage. A tidy-up pruning to keep it looking good whenever needed, especially after winter and new growth soon filled any gaps. Excellent plant.

Eccremocarpus completely transforms a fence into a display.

Other climbers were planted. They had to be soft and gentle for passing visitors, no thorns allowed. These included *Schizophragma hydrangeoides* var. *concolor* 'Moonlight' with its silvery veined leaves and *Clematis armandii* (named for Pere Armand David, better known for the wonderful *Davidia*) a rampant evergreen plant with white flowers each spring. *Lonicera henryi* 'Copper Beauty' (named for the famous Irish plantsman Augustine Henry) soon became a rampant thug. Its wonderful young bronze leaves arched over into the path of visitors, giving a great effect and only being pruned back when it became really hefty. This cultivar is a recent selection from a Dutch nursery, as was noted by the nursery owner while on a tour around the zoo.

Verdant summer growth hides visitors almost completely. Gorilla island is to the right.

An unexpected, borrowed landscape. The water in the far distance is actually the lake, but from this angle you would think the moat goes far further.

The overall impression as visitors walk around is one of verdant growth, large leaves, with large batches of colour, and limited views – as you would expect in a tropical rainforest.

The visitor stand-off planting changes during the seasons. Here *Geranium palmatum* dominates. A little later, *Inula magnifica* would be a mass of yellow.

Even in late winter there is structure and some colour, purple *Phormium* and *Ligularia* 'Britt Marie Crawford,' and silvery *Cynara*.

Asian lions. The landscape always looks bare to start with. After only a few weeks, the hawthorns were leafing out, helping to screen the rear fences.

Asian lions. What a difference three years makes. The silvery foliage of the *Hippophae* make this look very different. Further back, the dense growth of the hawthorns is working well to screen the fences. Gentle formative pruning, raising the crown of the *Hippophae* will ensure good viewing yet still keep the desired appearance and screen.

Sumatran tigers. Note 'hot vine' around central tree to stop climbing. Large clumps of bamboo from the zoo nursery were carefully sited to hide views of gates from viewing areas. Urban tree soil 'mulch' was spread over the entire soil surface. Odd clumps of *Carex pendula* break up the sides of the dry stream and pond.

Sumatran Tigers. After a few months, the RTF grass had formed a good enough sward, the patches make it more natural. Slow establishment was expected as high temperatures are needed for germination. The bamboo at the back is just starting to screen the fences.

What a difference four years makes...spot the tiger. The grass was very well established. There was very little mud still, except near the pools where water overflowed. The cuttings of aspen *Populus tremula* did extremely well as quick cover and are easily pruned if needs be.

CHAPTER 12

Sea Lion Cove and Flamingo Lagoon
Problems Working with Water

For many years, the flamingos had been on the main lake, fully open to the sky. This was great for visitors viewing the birds but not good for breeding season when seagulls would come down and try to take eggs or chicks and cause much disturbance at feeding time – none of which was good for the flamingos. The only remedy would be to construct an aviary to accommodate them. This meant working in and around the lake. Working near water was never easy, and there was always a list of possible complications.

The old flamingo habitat had taught the horticultural team a lot about water-edge plants and planting techniques, where erosion was successfully controlled with a soft edge of plants.

Several large trees, two with serious fungal problems, were removed, leaving this as the area for the new flamingo aviary, their original island is on the left. Any remaining trees were very important to screen the structure, once finished – builders please note.

Equally, the sea lions had been in a small round pool for years, which was far from ideal. The water needed much better filtration. Hence, a complete redesign of their habitat was needed too. Combining the two into one build made sense, especially as major changes were envisaged in the lake system. The old sea lion area would be filled in, and the sea lions given a new longer linear pool with lots of interesting artificial rock shapes in the water to make it more stimulating for them – enrichment of a different kind. To build this, the original small zoo lake would need to be partly filled in and dug deeper in other parts to facilitate the new sea lion pool. Staff reckoned this could become very interesting. Building a pool in a lake was never going to be straightforward.

As with all building projects, there were site meetings to see what problems the builders might meet, and the lake was an obvious issue. The lake is really a small stream that had been dammed many years ago. Imagine, lots of trees around the lake, a gentle flow from springs or surface drainage, and a metal barrier at the first zoo bridge near the flamingos (which kept geese on the big lake and ducks and pelicans on the small one). From experience, this was known to hold quite an amount of debris. One year, while trying to clear leaves from the barrier under the bridge, by chance, it had been discovered that it was actually hinged – three more taps, and it swung horizontal. The resulting flow took trailer loads of leaves and almost three gardeners with it. There was certainly a build-up of debris, sediment, or sludge – call it what you will. It was soft and gooey in the extreme, plus it stank when pulled out – this was proved over the years by the horticultural team as various work on the lake edges was carried out. The big question was, how much sludge was there? The considered opinion from the horticulture side there was at least a metre or two, maybe more. Samples were taken, which said far less. The less there, the cheaper the task, of course. Time would certainly tell.

Starting in late summer, a small earth dam was made near the bridge, separating the two lakes and the water was pumped out from the small one. Time was needed to dry out the sediment to make it easier to handle. The point was made that there was no flow from the upper lake at present, quite normal after a dry summer, but would it not be a good idea to build a little earth bank along one side to make a stream to take the overflow when it came? Assurances were made that the electric pump would cope. A small pump at that. Grand so. A month went past with a little rain. More rain followed, and the upper lake started to fill. Can the overflow be stopped? Yes, for a while anyway. Then it rained more and more. The point came when the overflow had to be allowed to flow. Displaying the usual courtesy and looking at the machinery in the way, the builders were informed of the need to let the water overflow. Grand, work away. The wee electric pump will take it…

Within half an hour, the water level rose enough to make the builders' feet wet. The pump did its best, but the level rose further until it flowed back over the sediment that had been drying for a month, ruining the work, and putting it all back a month. The very next day, a wee earth bank went in to make a stream to take the flow – in a good wet winter that could be an estimated one million litres a day (based on the surface area and amount the upper lake could drop), but it is hard to convince people at times.

Second draining of the small lake. Note the wee stream on this edge to allow flow. A sea of soft sediment awaits…

Second attempt. The sediment was dry enough to handle. A large excavator came in to make a deep hole that acted as a sump to drain it further. This worked but kept filling, so a larger pump went in to drain that sump as needed. Where could it pump to? It is only water – into the large lake? The excavation stirred up enough sediment to make the water much too thick to drink but still too thin to plough. As it was pumped into the next lake, a dark cloudy blanket spread downstream. Then the point was made about possible damage to any fish or wildfowl. It looked awful too. Into the drainage system it had to go.

The wee stream kept flowing all winter. The lake silt was replaced with stone and the concrete work started for the pool and surrounds. At the base of the tree, you can just make out a few horizontal lines. This was the original concrete edges to the small lake, completely buried by silt and vegetation and water – all now removed for a more natural stream-side planting.

Horticultural popularity levels fell a good bit with builders and engineers for a while, and fell even further when they realised they had not just a little sediment to remove, but 5,000 cubic metres of it. Too liquid to move by lorry, too organic for most dumps to take it. Normally, excavated lake silt is stored and left for a time – maybe a year or two to dry out and for air to get into it, to help the organic matter to break down, then it can be spread over land and ploughed in. Topsoil needs to be stripped off, a retaining bank made, and the silt dumped inside. This had been seen at Hampton Court on the Long Water, and it had taken two years before the sediment was dry enough to spread. Such material when still wet is highly dangerous. It looks dry on the surface, but if you fall through that crust, you can be in real trouble. A young school friend had done that in Twickenham during river dredging in Crane Park. He sunk into it up to chest deep, but his mum's reaction to him with caked-on mud from shoes almost to his chin was memorable, to put it mildly. Nor did we get thanked for pulling him out...

There was nowhere in the zoo that had room for that amount of sediment. There was a suggestion that the gardens team could spread it on the planted areas, between the plants. 'It is good for the soil,' was the comment. 'The Egyptians get great crops from the Nile floods and the fresh silt deposited...' This was not fresh silt! Can you imagine applying 5,000m³ of lake mud as a mulch while still wet, when every cubic metre would weigh a tonne? On top of the nice weed-free mulch? On top of dormant bulbs and herbaceous plants? Definitely not a realistic idea, folks! Eventually, with no other option, it was put into a corner of the upper lake to make a new planting area. Though after it spread underwater – for about 30m or so judging from air bubbles coming up – there was not that large an area left to plant.

After only one summers growth, the area still looks thin, the stream planting was just established, but the conifers are starting to form a green background. Sea lion pool is on the right, underwater viewing under the roof.

The other problem the builders realised was that they were building the new sea lion pool below the lake level, below the normal water table. Building a pool in a lake could be interesting. This meant they were making a waterproof pool with concrete, partly below water level. There was a definite worry that when empty, the pool could potentially try to float. Of course it would not float but might crack under the strain and not be waterproof anymore. Dozens of earth anchors had to go in to stabilise the whole pool lining. Part of the concrete was for visitor pathways, which included some planted areas. There was a knock-on problem for the plants here, of course. Some of the planted areas were holes in that concrete, which rapidly filled with water up to water table level, about a half metre below ground level. Other planted areas were actually raised beds, made of lovely large flat sandstone rocks, sitting on the concrete. These rocks were very large and needed special grabs on the machinery to handle them safely. Watering was not allowed for – a slip up from the horticulture team. Usually, irrigation pipes would have been suggested as paths went in, but this concrete went in so early, the opportunity was missed. Watering would have to be by hand as needed, though not too often as they are large soil masses, and this is not usually an issue in the average Irish summer either.

Drainage for these raised beds would be an issue. Even after rain, the beds would take days to drain. Typically, a few simple holes in the concrete would suffice, but not here. A hole to allow drainage would work both ways, wouldn't it? Water would be able to come up from below ground, as well as water going down? So, there would be a potential seep of rainwater or irrigation water (if a dry summer) from the raised beds onto the paths. The proof of this was a few minor seeps of water where two concrete areas met – different concrete pours on different days. If not sealed properly, there could be a gentle flow. Not a problem, until algae grow in summer, making it slippery or winter frost causes ice, making it more so.

One more possible problem was that the underwater viewing area planned was the lowest point on the paths. During rain there is a perfectly natural gravity flow downhill. It was very soon realised that there would need to be a sump at the underwater viewing area, with a pump that would take rainwater and pump it into the lake instead. This was easily done once allowed for early enough in the design.

The natural and varying lake levels gave problems. The initial design had no proper retaining barrier for the stream. The desire was for the planted earth bank, in places a mere 300mm wide, to contain any water. The worry was that with a high lake level in winter, if it overtopped at any point, the flow would erode the soil bank, trickle (or flood) down to the sump pump at the underwater viewing and the flow would overtake the pump. A bigger worry was that the lake level was fully dependant on rainfall and groundwater spring flow. There were no measuring devices on the inflows at all. The level was maintained by gauging the flow and the level by eye, and estimating by experience the possible extra height the lake level would reach. It used to be tricky – too high and paths flooded, too low and islands became mainland – not good for some of the primates. Now, if the lake level went too high, the entire underwater viewing area would really be underwater – and very accurately named, as someone kindly pointed out.

After much discussion, a retaining concrete wall was added along the stream as a base to the usual stand-off fence. The fence is an unfortunate essential with water, as the area would look far more natural without it. This wee concrete base wall would hold about 200mm of extra lake level. But at one point, the path dipped and caused a puddle, which the builders solved very easily by drilling a hole through the concrete from the path to the lake. This was grand until the lake rose enough. The drainage hole did work – but both ways. It was also effective from the lake to the path... The underwater viewing area had a nice paddling spot for visitors until the 'drain' was plugged, and another solution was found. Water is its own master. It is never easy to hold back if it wants to get somewhere. Drama is guaranteed.

Back to design. What effect is being aimed for? What is the inspiration for the sea lion habitat? The Northwest American coastline. Think of a beach on which the sea lions haul out at times. 100m wide shoreline. Think of sand dunes, maybe another 300m. Behind that would be a coniferous temperate rainforest (typical of Northwest America) for as far as you can see. Simple. Of course, there was only a maximum width of perhaps 20m to accomplish this, as the visitor path followed the stream that was the flow from the upper lake to the lower lake. 'Stream' is stretching the definition of the word a good amount as it is only a stream when the upper lake is full and overflowing, and this usually only occurs in winter. In many a dry summer, there would be no flow at all – just a thin ribbon of water left. The planting would be dense enough to screen the lack of water in some areas.

Elements of the natural habitat were included at planning. Not least, the timber totem poles, carefully researched and very beautifully carved, giving this a unique North American First Nation appearance. A sand dune with the skull of a whale 'washed up' is really eye-catching. Planted with lyme grass *Leymus arenarius* (from the Greek *elymus* meaning cereal and *arenarius* meaning sandy), a sand-loving grass that had been used elsewhere before but never did that well. Well, sand-loving proved right. A lorryload or two of sharp grit and sand (the same as used in the elephant habitat) made up the beach. It was planted up, and within eighteen months, some of the grass had to be pruned down as the whale skull, 1m tall and 4m long, could hardly be seen. Regular weedkilling, or rather 'editing the plant,' now controls its spread. It is often used for reclamation work or sand dune establishment. It is incredibly effective, binding the sand grains to prevent erosion, yet allowing other plants to get established. Very useful. But in a 'garden' situation, it is an absolute thug if happy, but very attractive with silvery leaves, it does make a good show around the skull. A couple of *Hippophae rhamnoides* left over from the lions look well, especially covered in the bright orange berries and, of course, looked very natural in the sand dunes too.

A very different feature to set into a naturalistic planting. Totem poles lend an authentic feel too.

The whale skull has a history, of course. In the mid-1990s, various ideas had been tried to increase visitor attendance – a dinosaur exhibit was first, then a whale exhibit. They obviously used models, most half sized as many whales are soooo big. The skull had been donated from Trinity College, Dublin. It had been stored for years there and was as green as a good lawn – bless the algae as pioneer plants. A thorough clean before it went on show, and lots of people could not believe that the visible 200mm hole at the end of the skull is actually the spinal cord access, not the throat. The following year, the skull was stored near the old nursery, and a few seedlings popped up from cracks in it. They looked different, so the questions began. A balsam *Impatiens* of some sort, but which? When it flowered, it was identified as small balsam *I. parviflora*. Not a common plant in Ireland at all, so how did it get there? Looking up references in the excellent 'Flora of County Dublin' by the Dublin Naturalists Field Club, it was only listed in two places. One was a damp lane by the zoology and biochemistry buildings

disturbance of the birds, so no viewing from two sides. With a little formative pruning and regular annual sessions, this will form a dense solid screen – perfect for the flamingos' needs.

The *Podocarpus* was an almost instant screen, fully blocking the view as desired for the flamingo's benefit – the five freedoms again. It would need a form of pruning later, best described as 'chunk removal,' or as a colleague in another zoo described it, like a cow eating a piece of it. The idea would be to keep the screen, but limit width and height, hence, keep it dense. *Darmera peltata* was planted below it, giving interest. It flowers in spring from bare soil. There were comments galore that the *Gunnera* is seeding around…

Coast redwood *Sequoia sempervirens* had to be planted, though obviously none of the four 2.5m specimens planted will ever get to their full 100m height. It immediately provided a very different appearance, the dark green scaly leaves having a very different look. *Sequoia* is a most unusual conifer, apart from being the tallest tree species. Living near the coast, there are often dense fogs. The water vapour from the fogs can be absorbed directly into the leaves and through the bark, which helps the tall trees get water up high. If condensing onto the leaves, the drips that fall can give an estimated 25% of the tree's water needs – crucial in a dry summer. The roots can get flooded in heavy rainfall areas, with silt and debris deposited on the root plate. Often this is a cause of death for many trees, but not this one. New roots immediately grow upwards into any silt, and eventually, the old roots die off. With thick bark, fire is not a problem to an old tree, very beneficial as the fire will often reduce competition from smaller trees and shrubs – a very useful strategy on nutrient-poor soil (due to heavy rainfall). One last trick. *Sequoia* is able to regenerate from cut or broken off trunks after storm damage or even felling – new shoots will arise from the roots.

Eastern arborvitae *Thuja occidentalis* 'Brabant' is another mass-produced hedging cultivar.

83 root-balled specimens of western red cedar *Thuja plicata* 'Martin' were bought in at 2m tall. They gave an instant effect, but as above, they were all identical. They are a very common cultivar and excellent for a hedge. Within a year or two, with no annual pruning, the formality wore off. Interestingly, a visitor said they did not like seeing so many very common hedging conifers being planted, and that it was not like our normal planting. It was an excellent opportunity for the team to explain why they were being planted and how the plants create the right habitat appearance, but it is good to hear that the planting is being noticed. Various tribes within the indigenous peoples of the northwest coast of the Pacific used this tree for totem poles, as it is very resistant to rotting – the average totem pole may last 75 years.

Western hemlock *Tsuga heterophylla* is a rapid-growing tree. It grows up to 1m a year, if happy. Four 2.5m specimens went in, giving an immediately different appearance.

In nature, under the conifers, there would be an understorey of shrubs, wherever light is good enough. This understorey gave a much more colourful range of plants to choose from; several from the correct geographical regions. Equally, it was good to have a break from the coniferous green everywhere – the same principle as in the Kaziranga Trail where the mass of bamboo needed some breaks with trees or shrubs. Looking ahead, there was the upcoming project on a new wolf habitat. That would involve a raised walkway, retaining gabion walls, and a need for dense evergreen screens to hide it all – hide visitors too – and preferably, stay in keeping with the planting theme for the sea lions. A selection of shrubs, including some evergreen ones, would therefore be needed too, if possible.

Flowering currant *Ribes sanguineum* is a well-known shrub, often planted in Europe, from western North America, so it was correct for the themed area. It is readily available in good sizes and cheap if bought as bare-root plants in

Here is a good place to mention a method of planting on water edges. Experience of several years trying to stop soil erosion in the old flamingo habitat proved very useful. The problem there had been to get the plants established as the flamingos walked over the plants when small, causing a lot of damage. Taking bundles of twigs and laying them down in marshy ground is a very old way of creating a pathway. The twigs spread the weight of the walker. Very simple and effective. If kept constantly underwater, the twigs will not rot for a long time. These simple bundles, usually called faggots (but not to be confused with the meatballs of that name, famed in parts of mainly northern England), are cut from coppiced stands of willow *Salix* or birch *Betula* and would have been used an awful lot historically, even as fuel. Coppice wood was smaller in diameter, hence much easier to cut by hand. Poor families without an oven could take a dried bundle to the baker as payment, and he would cook their loaf for them. Ovens then were simple brick or stone. The fire made inside was similar to a proper pizza oven nowadays, and the loaf was cooked after the fire died down enough. The bottom of the loaf often became covered in ash. You only ate the top half if you had money – hence, the term 'upper crust.'

For many waterside applications, willow *Salix* is used. Often available nearby as it loves wet soil. But – and it is a big but – it is very happy to grow from cuttings at any time of the year unless completely dried out first. For our purposes, birch *Betula* was far better as it will never root. It has many more side shoots making it bushier, and better still, there was plenty available – if needs be by quietly raiding the browse feed that was delivered daily for some of the animals.

Several large faggots, about 200mm thick and maybe 2m long, had been prepared for the old flamingo islands. The sneaky part was the careful placing of the plants within the faggot while being made – tried and trusted *Carex pendula* and *Iris pseudoacorus* again. At the flamingos, they were lined out just at water level and staked in position using willow. These were cuttings, maybe 25mm thick, and would certainly grow. Some shrubby growth would help shield the flamingos, and the willow roots would stop erosion. Excess top growth could be pruned down as needed. It worked brilliantly. There was no erosion, better screening, better breeding success too.

With that experience from ten years before, more faggots were used along our stream, but with variation, as there was no wildfowl in large numbers to worry about. Instead of placing the plants within the faggot as it was made, planting could be done behind it on the edge itself – much easier and quicker. The amount of twiggy material was less as no plants were bound within. This made them quicker and easier to make and handle.

The stream edge needed some form of erosion protection for a while. Simple thin birch faggots came in useful and were both easier and quicker to plant behind. *Iris pseudoacorus* and *Caltha palustris* form a planted edge, while groundcover *Fragaria* rapidly covered the soil and reduced slippage.

Same idea but a variation on the theme, using *Ajuga reptans* 'Catlin's Giant,' another rampant grower, that will reduce surface mulch slippage and *Caltha palustris*. The twiggy faggots are disappearing fast.

For the new flamingo habitat itself, a lot of larger, heavier faggots were needed again, with a range of plants in them. The usual principle of if one species fails, another may do better. The same faggot method was used, with the bundles forced tight by ropes and a natty foot operated lever system and tied with thick hemp string for security (so everything can rot eventually, no nasty bits for flamingos to get tangled in). For contented nesting, the birds needed a water-surrounded island, with all the usual soil erosion issues on the edge. Clay rich subsoil had been brought in with the hope it would resist erosion. The usual pendulous sedge *Carex pendula* would always do well but gave just a green background. Yellow flag *Iris pseudoacorus* was a guaranteed plant with a great display of bright yellow flowers in early summer, but the drawback was that they could become too tall and screen too much. The giant marsh marigold *Caltha polypetala*, which had been so good in the gorilla habitat and on other islands, had to go in. Again, it grew well but not too high and also looks good from a distance. Lesser spearwort *Ranunculus flammula* also looks good. This is a typical yellow buttercup flower with a creeping root that loves water. With a few willow stakes, allowed to grow a little but easily cut down if needed, the lakeside edges of the island were secure. Not surprisingly, any edges not planted along the rear of the nesting islands did suffer erosion caused by the birds getting in and out of the water. The same planting mix was used along other edges in the habitat too. Lesser bulrush *Typha angustifolia*, which at a possible 2m was too tall for some areas, is a fine plant with narrower leaves than the greater bulrush *Typha latifolia*.

A day after the flamingos went back in, with temporary dams removed, the water level went back to normal. The faggots and plants are at the right level. Praise be. Note the extra twiggy birch through any planting, to prevent the flamingos walking through the new plants.

A summer's growth makes a great difference. All edges nicely greening up and giving the flamingos the visual barrier that helps them feel comfortable.

Two years later and looking very dense. The *Iris* has really established well, while the *Caltha* with a very different leaf, adds to the natural look. For many years, this was the best breeding flock in European zoos. Nice to think the plants played their part too.

With fewer waterfowl, there was a chance to try some water lilies *Nymphaea* again. This had been a request in the main lakes, but all attempts had failed due to damage from the waterfowl. One year, some vigorous deep-water plants were tried in the lower lake. The hardest part was trying to plant them. It can be done with ropes on either side of a pond, positioning where wanted, then letting go of one set of ropes and removing them. The lake was definitely a little wide for that, so the boat came out, and the pots were gently lowered down on a string. Then the problem was figuring out where we had planted them to place more nearby... We couldn't see a thing in disturbed water. Some bright spark said to tie the string to an empty plastic bottle as a marker or a buoy, leave it there until the job was finished, and then cut them all off. This worked well, apart from three pots where the string was a little short, and over lunch, they drifted downstream. They had to be chased and herded back to the right spot later. The wildfowl, no doubt coots, kept nibbling at any leaves that tried to grow, and eventually, they all died off. A few hefty older plants, donated from staff clearing out their garden ponds, were planted at the edges of the flamingos as it was too deep in the middle. Some have done well, spreading out a little as they get established.

The only waterlilies to do well in the zoo, protected from the general nibbling of ducks, geese, coots and moorhens.

There was one problem planting in the flamingos that had never been encountered when working on lake edges before – there was no water. The area had been drained by the builders to make working easier. The water was not due to go back in until all the work was finished, and that would be a few days yet. How do we get the right level then? Ask the builders. Sure enough, with a little training, we mastered their laser level and had the unique experience of marking planting points by laser level and marking paint. All plants in faggots were very well watered. Another unique experience was standing in a dry lake with a hosepipe watering marginal plants. Necessary, and it worked. The great day of the dam bursting came soon enough. Though, it was not too dramatic as the water levels were not that different. Amazingly gratifying to see the water creep up to and half-submerge the faggots. Perfect.

A corner that the animal teams did not want the flamingos to get into, well treed up, with many cuttings in too. Reliable *Carex pendula* and small plants of *Libertia chilensis*, another guaranteed plant. The faggot has small plants of *Iris pseudoacorus* and *Caltha polypetala*. Note the green line at left – the marker for water level while positioning the faggot.

With almost any Irish freshwater body, there are nearly always reeds *Phragmites australis*. It is an absolute thug of a plant, creeping over mud surfaces to spread and, in some places, as an invasive alien plant, causing a lot of problems as it is so hard to

The urban tree soil was laid to a depth of about 400mm. The membrane was laid over it, then the topsoil. Trees need to be planted. Wet soil, so a willow was appropriate, but something nice to look at even in winter. It would need to be vigorous – the aviary roof was only 12m, so any tree would need coppicing regularly. Coral bark willow *Salix alba* var. *vitellina* 'Britzensis' ('Chermesina') was chosen, mainly as it was readily available in good sizes. A few were ordered in, and then panic set in when people saw how hefty each main support post for the aviary was. Several more of various sizes were suddenly needed. Of course, they only hide the post when visitors are standing in exactly the right place, but they certainly do soften the look. As always, tree staking was needed, but as usual, 'discreetly' if possible.

Here that was impossible. The trees were bare-root, so cables over the root-ball could not work. Reducing the height of the tree helped a lot, as that made it less top-heavy. The stakes could be shorter, which was good as the subsoil was very hard to drive stakes into – it was the original lakebed, with loads of infill over the years, including lumps of bricks and concrete. All that pruned-off material resulted in a large number of cuttings, very handy and very sneaky. Instead of a 'tree,' each support pole ended up eventually with a copse of willow hiding it, two or three trees as supplied in different sizes, but also pruned down to very different heights, giving lots of cuttings. The great advantage being they were much easier to plant – just drive a steel bar through the membrane and push the cutting in. The base would be in the natural water table once the lake level returned to normal. Willow can be planted as 'truncheons,' 25mm thick rods that can be hammered in if needs be, then cleanly trimmed afterwards. Some of every sort were used to vary the appearance and heights as much as possible.

Once main tree planting was finished, planting the cuttings was the last job as they would be easy to put in. Then screened topsoil was spread about 50mm thick. The screened soil came in dry, so it was very easy to spread. It rained that night and next day… Wherever anyone walked, their boots picked up the soil until you had 200mm thick soles and twice as wide as usual. Of course, each footprint left a hole in the topsoil, down to the membrane. It was hard to explain to non-believers how it would bed down in time and hard to patch up the surface as it was constantly being walked on . Normal grass seed was sown but not expected to grow with all those footprints. Sure enough, it mainly failed, except for the odd edge and corner. After a year, there was an almost full growth on the soil, but not grass – literally. Knotgrass *Polygonum aviculare* is an annual, often seen as a weed in cultivated or waste ground. It must indeed be tough to survive so well here, almost a monoculture. It looks ok too.

Getting there… The new flamingo island being formed. Most trees already in to get established before growth, as all were bare root.

Any shoots pruned off the willows were kept for cutting material. They were very easily planted – a simple hole through the permeable membrane and push, the water table was about 250mm below soil level. Heavily mulched after planting to reduce weeds.

Plant growth took off rapidly with an almost constant water level and plenty of free fertiliser courtesy of the flamingos. The trees grew a couple of metres in their first year – some needed cutting down within eighteen months of planting. Keeping a natural look when regularly cutting down trees can be tricky and is best if cuts are near ground level. Coppicing some of each group of trees allowing regrowth from near ground level each year will give a succession of heights that will always look natural. Never, ever prune the top 2m off as it looks unnatural. It is far harder to prune at height and will grow back far quicker, and need pruning every year.

The stream in early summer, very full of growth. The massive leaves of the *Gunnera*, taller to the right as on a 2m mound. The distant silvery *Salix exigua* is an eye-catching highlight. The dark red *Rodgersia* flowers extend the colour season and in the foreground, a volunteer planting of *Myosotis*, no doubt having drifted down from the gorilla habitat as cuttings. Magic.

CHAPTER 13

Orangutan Forest
More Watery Challenges

Dublin Zoo has always benefitted from having a large body of water. There are always some distant views across the lake, especially on entry, and it gives a great feeling of space. Developments during the 1990s made several new islands in the lower lake, which took a good while to settle in, mainly as there were so few trees around that lake then. A similar worry arose when the African Plains was first developed. Initially, there was a plan to make an island near the entrance, which would have ruined the view as you entered. The lake's depth had not been factored in… Fortunately, all of the budget for the island went into making the entrance road wider. Plans for making a new island for the orangutans was worrying as more lake surface would disappear, giving a less open aspect. Besides, after the flamingo lagoon aviary had been built, there was a prime need to maintain some of the current island trees as screening.

Concerns were discussed with the design team, and the plans came back with good modifications. An existing island with mature trees on it would be partially left as is and animal-free. The trees on that island would still screen some buildings and frame the new island too. Another existing island would be extended lengthways but kept very thin – in places only 2m wide – forming a long screen at the rear of the orangutan island, giving a green background. More importantly, this kept a small group of *Eucalyptus* that screened the flamingo lagoon aviary from the front gate viewing area. It also screened visitors from each other on either side of the lake across the island – the standard design aspiration of avoiding cross views. The previous year, additional trees and shrubs had already been planted on the opposite lake edge. These were nearer to the visitor path and allowed an increase in the density of plants, plus the opportunity to vary the planting there.

The view as visitors entered the zoo circa 2015 is of a wide-open space, green and welcoming, with almost no man-made structures visible. The new orangutan island would be at the far end of this vista.

The plans showed modifications to an existing lakeside house for Siamang gibbon *Symphalangus syndactylus*. This was going to be a shared habitat, as it is good for both primates to have company, and it creates a much more realistic and interesting habitat for the visitors to see the animals in. Though more plant worries for the horticulture team as planting for two primate species would probably be more challenging...

This was the third 'new' orangutan habitat built in twenty years. The original had been a horrible square flat grass area with one large, dead tree that was unclimbable for the orangutans. The next one, created by moving the visitor path onto an island, was treble the area. It was still flat grass but with tripods of posts and ropes, now climbable. However, the orangutans had no reason to climb – the food was still served at ground level or back in the house, so why bother climbing? This time was different but still included the original habitat, which connected the new habitats to the house. The design considered the five freedoms of animal welfare again, particularly the freedom to express natural behaviour. Orangutans live in the trees. They move from tree to tree on thick lianas (a woody climbing plant that hangs from trees) or adjacent branches. They make nests in the trees by breaking branches and folding them in on themselves to form a platform. They drink from epiphytic plants growing on the trees, such as bromeliads, urn plants, which fill with rain. They can descend if they want to, but as fruit eaters, they find the most food up high in the trees.

Now think of being in a tree yourself, even if you cannot climb too well without a ladder. Can you sit on a branch comfortably when it is not that thick? Even a ladder can be painful on your knees if you lean against a rung too long. Often in zoos, the climbing structures are built using quite thin poles, easy to buy and transport, but seldom used a lot by orangutans. What was needed was a large tropical emergent tree, with a trunk around a metre thick and branches about 400mm wide to sit on comfortably and long vines strong enough to support an orangutan for access up and in between. The only option to recreate this was to make a concrete 'tree' and use ropes as 'vines.' None of this was horticulture of course, but great to see it develop and try to figure out any implications for the plants. The concrete 'trees' were planned as stumps only, with a few broken side branches, keeping the diameter thick.

The islands themselves were built using very large blocks of limestone as an edge of up to a metre long and wide of varying sizes, stacked three or four high. These rocks were firmly pressed into the lake bottom, reducing the risk of subsidence later, which had been a problem with 'new' islands previously. The island was then infilled with coarse stone and built up to the same height as the visitor path, allowing a much better viewing experience since you could look the orangutans in the eye when they were on the ground. It is always best practice not to be looking down at any animal – but this meant the visitors' view was of a large stone wall, around the islands, about 1.5m high, which would need softening.

Watching a large earthmover gently nudging the stones into place was always entertaining. The driver could almost make the machine talk. Once the walls were built and the coarse stone infill levelled, supplies of topsoil were put in place. For the new rear island and the old ones with trees on, this was the only time access was easy, while a teleporter could lift on bulk materials easily. Temporary access ramps for staff were essential. The teleporter was needed here to remove arisings – one old island had to be stripped of ivy *Hedera helix*. It is a wonderful wildlife plant when up on a tree stump and flowering, but an absolute thug on the ground. It swamps smaller shrubs far too easily and creates a boring monoculture of little use.

A huge advantage of having the teleporter onsite was that it could lift masses of tree chippings onto the islands as needed even before planting. A grab lorry delivering tree chippings from heaps elsewhere in the zoo filled the teleporter easily and quickly. All that needed to be done then was spread the chippings by hand. This saved so much time and the manual work of getting the tree chippings in place, and, of course, a deep mulch reduces weed growth and watering needs for any new planting. In relation to it being too dry, these planted islands were 1.5m off the lake level, almost a raised bed and with a coarse stone subsoil. In effect, they only had rainwater for irrigation. A deep mulch was used throughout, which was very useful for retaining soil moisture, especially since there was no easy way of watering once the orangutans had access.

The foundations for the concrete 'trees' went in. The engineers went overboard, but they had to. To hold up the 15m concrete trees, they specified a 5m square foundation, 2m deep, on top of which plants were expected to grow. After much discussion, the top foundation level was dropped to give a minimum of 250mm of topsoil depth. Only grass was expected to grow. The chief worry was dry weather and brown grass.

Limestone blocks create the new island, with massive concrete foundations built for each concrete tree. Look at the birch tree in foreground, a self-sown seedling, with a great natural bend and growing right on the edge having germinated no doubt among the *Carex* there. Note: behind the orangutan island, the distant views of other paths – and signs too. These all needed to be hidden – no cross views, hence the rear island for planting.

The concrete 'trees' were based on a heavy steel framework, with expanded metal secured as the basis for the concrete 'bark' to be applied. Large branches were finished on the ground and then teleported up and bolted and welded into place, all built to a design from specialist concrete artists. And they were artists. The finished effect is very realistic. Steel eyes were built in for rope attachments (the ropes are actually made for deep-sea trawler nets, usually hundreds of metres long). Inside some trees is an elevator for food – hopefully, to make the orangutans climb for food treats. One tree had no ropes attached, just a stand-alone stump. Within days, a seagull had nested there. At one end of the new island, behind the orangutan habitat, a dead tree was wedged, overhanging the water. It immediately became a favourite perch for even more seagulls.

The wide concrete branches, destined to be comfortable seats for the orangutans and Siamang gibbons, to be fixed when finished at 10m above ground level, had great detail that only the orangutan could appreciate.

Work in progress, view from front gate, builders site huts very obvious, as is the white frame of the flamingo aviary, hence the tree planting to hopefully screen it further.

Befriending the machinery drivers safeguarded the trees – here the *Eucalyptus* in the centre are growing on an existing island that will become the new rear island and they will help screen the flamingo aviary.

Once the islands were constructed, the first task for the horticulture team was to get the rock edges planted up. Now consider this task, especially with health and safety in mind. On a building site, work was going on overhead in places – large chunks of steel and concrete moving around and items possibly falling. Planting along the waterline proved impossible from soil level, 1.5m above. Rocks had been so well positioned that there were seldom handholds or footrests either. The water was just too deep for waders, potentially 2m deep near the rock work and impossible to judge the depth when the water was disturbed. The only option was by boat, while avoiding any work going on overhead at all times. Great fun.

What to plant? Pendulous sedge *Carex pendula* again – tough, water-loving, available around the zoo in various sizes. Yellow flag *Iris pseudoacorus* again – readily available as liners and some available around the zoo but not so easy to dig up from the lake edges. A new edge plant was the giant Japanese butterbur *Petasites japonicus* var. *giganteus*. This had proved very successful on the Kaziranga Trail in front of the bull elephant pool. If planted on the island, it would still be contained and not too much of a threat – though, its rampant empire-building tendency is always a wee worry. It was early summer, the original plant was in full growth nearby, but it was reckoned it would survive almost anything. Roots were dug up. Leaves were removed to reduce water needs. Chunks were tucked into spaces wherever they would fit, maybe halfway up the rocks – wherever there was some soil to bed them in. The *Iris* was easy enough, as small and tuck-in-able, but it was hard to find suitable spots for it as it had to be at lake level and in water.

The *Carex* was the hardest to plant. Dug up from the lake edge nearby, clumps were then wedged tightly into suitable spaces, near water level or in pockets between the rocks where there was some topsoil. Previously, such work had seen mute swans *Cygnus olor* trying to pull out the *Carex*, so these were really rammed home – the handle of a lump hammer can be very effective. Imagine coming back after lunch to find some *Carex* floating away… The swans were nesting and wanted a few leaves. They did not give up. Tugging hard at the newly-planted *Carex* they had managed to pull some out far too easily. Back to the drawing board, return with birch stakes, with maybe 40mm of a side branch sticking out as a peg and hammer them into the *Carex* clumps. Not always easy, those rocks resisted many angles but were successful by and large.

All this work was done from a small two-man rowing boat. This planting needed careful organising. Launch the boat, moor it, chained in place until ready – for safety, at a padlocked gate that opened onto the lake. Three staff had to be involved. Two could row out to the island, one to be marooned there. The other staff member rowed back, picked up the third person, ropes, stakes, hammers, watering can, etc., locking the gate on the lake edge from the boat. They could then row down the lake to pick up bags of plants just dug up by a fourth gardener. Finally, they would row back out to the island. They would throw ropes to the marooned staff member, whose job it was to keep the boat near the island. One rope was tied to each end of the boat to allow leeway. The rower then had to keep the boat far enough away from the island to enable the third man to plant and wedge, which was impossible if the boat got too close. The plants also had to be watered in – there's another unique one for the record, irrigating with a watering can from a boat. At least the water supply was handy. Slow and fiddly planting, but well worth it in the end.

Not everything survived, but enough did to make it worthwhile, and it certainly helps to soften the island edges. The effect should increase as the plants spread, especially the *Iris* and the *Petasites*, which potentially could cover large areas. A few spots were carefully designed nearer the viewing areas to allow more planting on the island side, and these have done well, mainly with the *Iris*, but great care was needed to ensure that there was enough of a clear gap between the island and shore at all times.

Not all the water-edge plants survived, there was so little soil or suitable places between the rocks, here *Carex pendula* is grand and will maybe seed in nearby too.

The *Petasites* has done very well where happy, not surprising, the large round leaves not touched by the orangutans or the gibbons. A bonus. In the background, *Inula* is not touched. The *Carex* is actively nibbled by the gibbons, the tip of each leaf grazed or pulled at to get a mouthful, limiting growth as the leaves are only half the normal length, but it will still survive.

Plant selection for the orangutan islands was very different. The rear island was designed to have no animals, merely a screen, so plant selection was easier. The overriding concern here was height. There was a risk of falling plants creating a bridge between the islands, to the shore, or worse – both. It was also required to be evergreen, so still effective in winter. A very deep layer of wood chips had already been placed on the topsoil as mulch, making it harder to plant, as did the narrowness of the island in places. Planting density was far too thick, an oft-repeated error, but a quick, dense screen was required. There will be casualties as plants crowd each other out, and pruning will be needed, which is not easy on such a narrow island.

Plants Used to Give A Very Varied Effect On The Island

Italian alder *Alnus cordata* is a wonderful tree, glossy green heart-shaped (cordate) leaves and quick growing. Available readily as whips, bare-root under 1m tall and therefore easily planted. Like all *Alnus*, the roots have a symbiotic relationship with nitrogen-producing bacteria, which form nodules on the roots, enabling better and faster growth. This species tolerates drier and poorer soil. It is often used on landscaping around buildings and is great as a windbreak – a good tough tree. Inevitably, they will get too large and may need cutting down. Hopefully, in the winter, as they will grow well again as a coppiced tree.

Japanese laurel *Aucuba japonica* 'Rozannie,' with purely green leaves (not as in the common awful, variegated variety), is a great tough shrub, not minding dry conditions. Tolerant of pruning if needs be, it will be a significant presence, especially as evergreen. The common name is misleading, as it is no relation to either the hedging common laurel *Prunus laurocerasus* or the bay laurel *Laurus nobilis* – the true laurel as in poet laureate or laurel crown or even cooking.

For a more tropical-looking evergreen screen and, of course, variety, some bamboo was needed. Square bamboo *Chimonobambusa quadrangularis* growing to about 2.5m, was planted in a few well-chosen spots. If the culm grew out at an angle, they might tempt the Siamang gibbons to try to jump off – not a good idea. If happy, this bamboo will form a dense screen, ordinarily straight and vertical, with leaves along most of the culm. Some more careful spots were selected for the walking stick bamboo *Chimonobambusa tumidissinoda*, about 2.5m tall. Doesn't the Latin name just trip off your tongue…after a bit of practice. This is a rampant spreader – rhizomes will push everywhere, and the culm will push up through other plants, but not long or strong enough to be a problem, hopefully. Indian fountain bamboo *Yushania anceps* has the most wonderful arching culm to about 3m. Planted at each end of the rear island, it should give a magnificent effect once established, especially from the front gate.

Hazel *Corylus avellana* was used merely as a temporary filler, cheap whips again, planted throughout the rear island, with the intention of removal if too large in a year or three and let the evergreens take over more. Quick growth is always useful at times to make a screen denser sooner.

The need to have an evergreen screen for the flamingo lagoon aviary further down the lake suggested some more *Eucalyptus* – there was already a few on one part of the island. They were not used much in the zoo as so obviously Australian for anyone who knows them, but here they would give that evergreen effect and quickly too. They needed careful planting, partly to prevent a possible leaning tree allowing primates to jump off; hence they were planted at the very ends of each island, but more so to plant them in the right place. The view from the front gate decking towards the aviary was all-important. By marking with canes and a flag, this could be checked. It was quite surprising how different it looked from the gate compared to first guesses. Check twice (or three times) and plant once. *Eucalyptus spp aggregata*, *delegatensis*, *gregsoniana*, and *pauciflora* subsp. *debeuzevillei*, were all planted in very small numbers. Time will tell which turns out the best.

The Japanese aralia *Fatsia japonica* has been much used around the zoo, mainly because many were bought in for the dinosaur exhibit, and once that finished, they were lifted and planted everywhere. This is a tough evergreen, with large 300mm wide palmate leaves that do look tropical. It makes a dense screen, can be pruned if needs be, and tolerates dry conditions – the perfect plant for here.

The rear island, well separated by water. No tree planting too near opposite edges and no water-edge planting either, to keep the island isolated. This is really the orangutan's view. Note the *Eucalyptus*, no real harm done despite the close work around some of the root area.

Griselinia littoralis, a cheap hedging plant as used in the gorilla habitat, also went in. Another good evergreen shrub and easily pruned, but sacrificial if needs be in years to come.

Bay *Laurus nobilis*, again as used in the gorilla habitat, was dotted through the rear island, another good evergreen and nicer than the *Griselinia*. It can get quite large – there are others around the zoo to 8m – which would also be good here, and it could be hard pruned to the ground if needs be. Very hard to kill once well established.

Himalayan honeysuckle *Leycesteria formosa* had to go in. Rampant growth quickly fills gaps and with the benefit that nothing eats it. And sacrificial too.

Delavays' magnolia *Magnolia delavayi* was a gamble. Some common names are not worth the effort, are they? A great evergreen that will look good even from a distance if they flower. Worries they may be swamped by neighbouring plants while still small, and they are far less amenable to being pruned, but worth a try. Great, if they work.

Foxglove tree *Paulownia tomentosa* was used in many areas already, with wonderful large leaves. If they do grow big enough to flower, they will be more visible here than elsewhere, as they are often flowering over peoples' heads above a path. *Paulownia* can be coppiced, but it needs light to grow afterwards. If the foliage around them gets too dense, the growth is too leggy and collapses easily.

New Zealand flax *Phormium tenax* is bombproof – apart from really cold winters of -15°C when they will suffer. A good dense screen at low level, but again carefully placed so the tall flowering shoots do not become a potential bridge.

Chinese necklace poplar *Populus lasiocarpa* has been used in a few places already, but a good tree and those large leaves will look great without too much competition.

Portuguese laurel *Prunus lusitanica* 'Angustifolia' is a much nicer form of laurel for hedging. If allowed to grow, it will become a small evergreen tree eventually, and it will tolerate pruning.

Caucasian wingnut *Pterocarya fraxinifolia* has the potential to become too big and was only planted in very carefully considered spots. It suckers too, forming a dense clump after many years. The large pinnate leaves will look good over the water, especially in summer when the pendulous flowers are present.

Chusan palm *Trachycarpus fortunei* has been planted in various sizes in several areas around the orangutans. There is a good educational handle on this; in the wild, one of the greatest threats to orangutans is habitat destruction to make palm oil plantations. This is not an oil palm, but they do look a little similar. On this island, only the smallest plants were planted. It was just too awkward to safely handle large potted plants on a 2m wide island...

Laurustinus *Viburnum tinus* is another common evergreen shrub, well suited to screening and easily obtainable, and as with almost all the planting here, it will tolerate heavy pruning if necessary.

From the other side of the lake, the rear island is simply dense vegetation, screening any views of visitors on the opposite shore and the orangutans – unless up in the trees. Growth after only two summers, with no gardening needed, weeding or watering. Note the white line just above water level, the lake has dropped about 150mm in a dry summer.

The orangutan habitat itself was a totally different planting scheme. A few plants were deliberately the same as the rear island, the idea being that it would merge the planting from a distance and make the orangutan area look larger. Many of the plants were copied unashamedly from the gorilla habitat as they had worked so well there. Orangutans are fruit eaters – more than 300 different types of fruit are documented in the wild, but they will eat leaves and bark as well, so the plants did have worries. The island is shared with the Siamang gibbon who has a different diet. Half is fruit, and the rest is flowers with a lot of leaves – 160 different food plants have been documented in the wild as having leaves eaten. The plants certainly had a few worries.

There were two areas to plant, the new island and the old original habitat. The new island was totally bare except for three small trees, one *Eucalyptus* and two sycamore *Acer pseudoplatanus* that had been on the original island there. The whole island was deeply mulched with wood chips from tree surgeons. This helped in many ways. There was still a lot of building activity, including a lot of foot traffic, and the mulch stopped the topsoil from getting compacted and needing digging over. At one point, a large excavator had to access the island again and though there were slight depressions where the steel tracks had driven, there was little actual compaction. Quite amazing what 200mm of chippings could cushion. The second area to plant up was the orangutans' old habitat – a bare grass area, needing few plants as it's so small.

As with the gorillas, many of the plants used had defences such as thorns, scent, hairy leaves, or, from experience elsewhere, had been not palatable to other animals. The different reaction from the two primates was quite fascinating, different eating habits, desire lines, and dexterity. The Siamang gibbons have smaller, nimble fingers and were even seen pulling weeds out of a plant that they were not eating and eating the weeds. Imagine willow herb *Epilobium* growing in a nursery pot with a shrub, planted out, the weed grows better. The gibbons were targeting the *Epilobium*, grubbing out the buds at soil level, and leaving the shrub alone. They actually graze the grass a good bit too, pulling off leaves in handfuls.

The main island was planted up mainly at the rear, so the primates could not hide easily from visitors unless they made a path. The area in front was grassed down, using the same mix as the gorilla habitat – some RTF Rhizomatous Tall Fescue with its deep roots and tough habit and a more normal rye mix for quicker establishment and appearance, plus some hay meadow wildflowers for variety. The desired effect was rough meadow, not a mown level 'lawn.'

Thorny plants were deliberately used a lot, trying to make the orangutans not use certain areas; for instance, behind the concrete trees where they could hide completely. Barberries *Berberis darwinii* and *B. julianae* were an obvious choice but only available in normal nursery sizes, a 3lt pot, and less than 300mm tall. There is always a balancing act between using small normal stock and sourcing larger specimens, which would need more care after planting, watering, etc. Unfortunately, no amount of asking would convince the primates to help with that. The barberries were interesting. They were left alone by the orangutans. The Siamang gibbons slowly, one leaf at a time, stripped the leaves off *B. julianae*, but not *B. darwinii*, which they left alone – this had been left alone by the gorillas, except for the flowers, remember. *Pyracantha cvs* planted en masse to reduce climbing in the existing trees were debarked but regrew. Eventually, over the years, they will probably fail unless they get a chance to get really woody. *Rosa rugosa*, largely left alone on the chimps' island in the plains, was successful here and good as it flowered. The flowers would be a good snack for the gibbons once they realise they are edible, but this behaviour has not yet been seen.

Thorns are not an assured defence! Here *Pyracantha* has been debarked, probably by the gibbons, but new growth is evident, maybe... Note though that the *Laurus, Prunus laurocerasus* and *Griselinia* are untouched, a welcome sign a year after planting. Of course, the buttercups are left alone...

Bamboo *Chimonobambusa tumidissinoda* was specifically requested by the animal team as the primates' love eating bamboo. They did indeed, and it lasted about a week – very expensive browse. Older tougher clumps of *Pseudosasa japonica*, a real thug of a bamboo, lifted from elsewhere and potted for a year or three, survived a little longer – the gibbons really enjoyed trying to get the leaves off. Strongly scented *Choisya ternata* had worked in the gorillas and it was fully left alone here. Again, it will be interesting if the gibbons eventually eat the flowers. *Inula magnifica*, not even eaten by geese, had been tried and left alone in the gorillas, but was swamped by grasses. Here it has romped away. Maybe the gibbons keep eating the grass and restricting its growth, or the grass here is weaker as it was sown on top of the deep mulch, so less well-rooted. Whichever, there are now massive clumps of the *Inula* and a very distinctive plant it is. The large leaves look good in summer, the yellow flowers look great and the dead flower stems add structure in winter. Incredible value.

Purple willow *Salix purpurea*, so good in the gorilla habitat, was also part of the shrub planting and quickly grew to 3m. Equally successful and entirely left alone. Time will tell if the primates here find the flowers as edible as the gorillas and the mangabeys did.

One of the best plants has been globe artichoke *Cynara cardunculus*, a massive plant when growing well. The gorillas had largely left it alone, so many more were used here with the same effect – neither of the primates touched it the first year at all, apart from passing damage, which is always expected. The large leaves formed great domes on all parts of the island. To pass by the orangutans went around or behind them. The second year the *Cynara* flowered, throwing up 2m tall stems. Marvellous to see so many and so little damage. Then, checking the island one day while the animal team was working there, little heaps of *Cynara* leaves were seen in two or three places. When asked why they were there, the animal staff replied that the orangutans were collecting them and using them as bedding. Tearing each leaf off the flower's stem, they made a heap with four or five leaves. Enough to insulate a little from wet or cold soil perhaps? Wonderful enrichment, great to see and no real harm to the *Cynara*.

The *Cynara* did very well indeed...almost too well in places, only a pathway left between plants. A very different look to the island though and a tough plant too.

In full flower the *Cynara* look terrific. *Iris pseudoacorus* have filled in well at the main viewing area, partially screening the limestone blocks.

Bay laurel *Laurus nobilis*, not badly touched by the gorillas, has done well. Evergreen, so this is forming a good screen in the habitat itself. Yellow flag *Iris pseudoacorus* did well as expected, especially in any constructed spots in front of the rock wall. When in flower, they almost hide the wall in places. Perfect. No damage from the primates either, neither of them like to go into the water, but the *Iris* is not palatable either.

Those concrete trees were crying out for climbers. The ropes were for the primates. In some spots, the concrete would be a great surface for self-clinging plants, anything except ivy. It was another gamble, as it often is, to see which plants would not be eaten or pulled off in play. Evergreen climbing hydrangeas *Hydrangea seemanii*, *H. serratifolia*, and *Schizophragma hydrangeoides* are all vigorous and with adventitious roots that grip really well. One plant of *H. serratifolia* seems to be doing very well and has not been touched. Many of these climbing hydrangeas do not appreciate disturbance and take a while to get established. Lack of water and weeding around them has not helped growth, but if one grows, that will be a bonus. Ivy of Uruguay *Cissus striata*, a form of vine, has proved very good at the South American House, not touched at all by the squirrel monkeys, which can be very destructive. *Cissus* really needs something to scramble into and through, so it will not be as good, but it still covers some of the buttress roots.

After four years, one *Hydrangea serratifolia* is doing well on a not easy to reach part of the concrete tree. Great potential if it is left alone, the trees would look far better if covered in vegetation.

In discussions about planting, the main desire was to have the orangutans backed by vegetation if they were on the ground. All planting on the main habitat was in the back third of the island, with only a few low-growing shrubs or some of the big herbaceous plants anywhere near the front. When asked, the animal staffs' opinion of plant survival was 'maybe a third.' The orangutans had a reputation for damaging both plants and fittings. Imagine the surprise when so little damage happened. In places, the orangutans only have a narrow path left to walk. They are surrounded by plants, most of which they will not eat and have not damaged – yet. The gibbons have been far more destructive with their extensive interest in leaves, but the variety of plants used has allowed enough to survive to look good.

When the primates were first allowed onto the island, the gibbons went first. They had to learn where to go, how they could get there, rope positions, etc and how to gain access to their house – a small door that an orangutan would not fit through. The orangutans went out after a week or so, on their own to start with. There was great relief when they actually climbed the ropes (which pass over the heads of the visitors at one point) to go down the other side for food that was carefully thrown in by animal staff. Routine came into play. There was more relief when they climbed back to go into their house later that day. When the two primate species met, there was interest. The orangutans and Siamang gibbons had not shared a habitat before and had to get used to each other. The nimbler gibbons sometimes riled the orangutans and were chased back to their house. Unable to fit through the Siamang gibbons' door and thoroughly frustrated, the orangutans pulled up the nearest plant and threw it through the doorway at the gibbons. A few plants suffered in this way, but not too many.

The gibbons spend much time seeking out edible plants, great enrichment. Unfortunately, almost always impossible to see what they are actually picking…

For visitors the sight of the orangutans crossing the pathway on overhead cables is always a great pleasure. Normally tempted by breakfast served al fresco on the island.

The original island, where the orangutans had been for years, had been planted with a similar mix of low-growing shrubs and seeded with the same grass and wildflower mix. Well, that did not go down well with the orangutans. They must have wondered who authorised this? Every plant was pulled up and thrown out. Not a single plant was left after a month or so. The orangutans just would not tolerate any new planting, though seeded in wildflowers are making a difference. Needless to say, that could be a problem if ever new plants are needed on their 'new' island. A problem for another day, but thoughts would be for cuttings of perhaps purple willow *Salix purpurea*, planted during winter and pushed down to within 30mm of soil level to make the cuttings hard to see. The same should work with *Leycesteria formosa*. Perhaps roots of *Cynara* or *Inula* could be de-leafed and planted a little deep. There is nearly always a way.

After three years growth, the concrete trees blend in much more with the newly landscaped areas. This is the view from the entrance of Dublin Zoo, taken in July 2019. Most visible planting is on the rear island, not accessed by the orangutans. Tallest trees on right are the *Eucalyptus*, carefully kept as screen to flamingo aviary – which cannot be seen. Perfect.

Apart from the main viewing area, there are a couple of smaller places where the animals may be glimpsed, but between visitor-side planting, looking far more natural.

The smaller original island was stripped of all new plantings by the orangutan, but the meadow seed mix is growing well.

There was not much new visitor-side planting, but one area was replanted with a mix of the palm *Trachycarpus fortunei*, the hardy banana *Musa basjoo*, yellow *Wachendorfia thrysiflora*, the wonderful large leaves of *Farfugium japonicum* var. *giganteum* in the foreground and a background of various bamboos.

Even on a cold misty day in winter there is still enough evergreen foliage to keep a screen and the wonderful leaves of the *Cynara* catch the eye.

CHAPTER 14

Soil Compaction and Urban Tree Soil

The answer lies in the soil

Topsoil, urban tree soil, and soil compaction have been mentioned and briefly discussed a few times already. Still, it is such an important topic it is worth having a separate chapter and detailing as much as possible about it.

Soil – such a simple, short word. Many non-gardeners would consider it simply dirt. However, it is the basis for almost all plant growth. The structure varies infinitely according to the ratio of clay, silt, sand, grit, gravel, and stones in it, the source of those mineral elements, which rock, acid, or alkaline, the plant growth it sustains, the annual rainfall, and the degree of drainage. An average soil may be only 250mm deep, under which there will be a subsoil that is equally varied, from solid rock of many sorts to deep sand, which also influences the topsoil. Healthy soil will be full of fibrous plant roots absorbing water and nutrients for plant growth concentrated in that top 250mm, with larger roots going deeper if they can for anchorage, which will depend on the nature of the subsoil. Desert trees and shrubs can go very deep for water, following cracks in rocks for many metres as water is the most limited resource they need in a dry landscape.

The soil feeds on the plant debris – dead leaves and wood from above, dead roots, etc. There is a vast number of microorganisms, such as bacteria and fungi, algae, and protozoa. Within one group of bacteria, the actinomycetes, there are many that are able to produce antibiotics useful for us, such as streptomycin. One bacterium *Mycobacterium vaccae*, may result in a rise in serotonin in humans if inhaled or absorbed (maybe via a scratch or even just with skin contact). A natural antidepressant – no wonder gardeners are generally a happy lot. A gram of soil will have countless thousand species of microorganism living in it – estimates are up to 25,000 species, but it's probably more as researchers find new species wherever they look. Actual numbers of bacteria per gram are in the billions. Some will help fix nitrogen, many will feed on carbon emitted from the plant roots, while some will try to cause damage. There is a constant battle between good and bad, and evidence that the plant can call for help by emitting various hormones.

Slowly go up in scale, and there will be larger organisms until small insects invisible to the naked eye will appear. Lift an old log or sift through leaf litter, and there should be a scurrying of visible life galore. Worms are famous for their contribution to soil health, making tunnels for air and water exchange, pulling organic waste into the soil to eat, and generally recycling all that lovely organic matter. There are thriving businesses selling worm compost, collected from cultured worms (i.e. cultivated, not educated…) raised on organic waste. Nurturing the soil is one of the most important tasks for anyone growing plants. Feed the soil, not the plant. Mulching (see below) with tonnes of organic matter helps feed this mass of fungi and bacteria. Soil erosion is one of the biggest problems around the world, with an annual loss far more than is sustainable.

So, soil is really important. It demands respect. Handling it when wet should be classed as a criminal offence. Destroy the structure of the soil, reduce or prevent water or air movement, and problems will follow. Now, imagine what happens to the soil if you walk on it. If dry, there will be some firming of the soil, which will sometimes not be a problem. If wet, the soil is softer and will firm more. If walked on regularly in the same spot, much more firming occurs until the top surface becomes compacted into a solid layer that prevents water getting through, so the soil then holds water and puddles form. Air is also prevented from entry. With more walking, a larger area becomes compacted and puddled. What happens then? The soil starts to die, it is less biologically activity, and, far worse for gardeners, roots may start to die as they need air to function. Eventually, entire plants – even mature trees – will give up the struggle.

What Is Compaction?

What is compaction? Why does it happen? Imagine all those different sized soil particles – clay, silt, sand, grit, stones, organic matter, etc. The finer the particle, the poorer the drainage, and the easier compaction will occur. The better drained, the slower the problem arises. Sand will take a lot of traffic before having problems. Clay drains so slowly it will obviously get very wet, very quickly. Healthy soil will have lumps and bumps, tunnels left from worms or dead roots, and water will find its way around roots. There are always cracks, often from dry spells that allow air and water exchange. That all adds up to good natural drainage. But put feet (or wheels) on wet soil and that structure is damaged. Then damaged more, until the soil becomes muddy, then wetter, even more muddy, then walked or driven on again, resulting in compaction. Many animals, including people, will try to avoid wet, muddy conditions, making more areas muddy. The problem spreads.

There is one other source of compaction that gets forgotten – rain. Bare soil can become beaten down, flattened even, by really heavy rain. This is especially so for newly spread and loose topsoil when it will absorb the water far quicker, rather than let it simply drain through. All those natural cracks and drainage holes have been destroyed by moving the soil – or digging it over. Once that starts, great care is needed, as a little traffic will cause severe compaction very quickly. Much better to keep the soil covered in plants, and there is less risk of erosion. The horticulture team started a habit of mulching topsoil immediately as a matter of course, rather than only after planting. This was usually with zoo mulch but sometimes, as mentioned with the orangutan islands, with wood chips from tree surgeons. This immediately prevented any compaction from builder's – or gardener's – feet and worked incredibly well. But it also stopped compaction from rain. Extraordinarily, it kept the soil from getting too wet. That does not seem to make sense, but it is true. Even after heavy rain, planting could proceed. No walking on wet soil – much easier, no muddy boots. The soil was remarkably 'normal' under the mulch, and good for working, plus the act of planting did not compact the soil either. The only downside was the need to pull back the mulch to dig and replace after planting, with the inevitable contamination of some mulch with a little soil and weed seeds. Though, the benefits outweighed the problems. If mulch is put on top of dry soil, then it will need more rain to penetrate – another possible downside but one that seldom occurs in Irelands' wonderful moist climate.

Now consider the cause of the compaction. It is usually feet. Some feet are broad – a gardener's average size nine boot, for instance. Some animals' feet are cloven and will push into the soil for a few inches, others are broad and flat and compress all over. A light foot will still cause compaction, but obviously less or more slowly, than a heavy one. Famously, sheep are the very worst or best at compacting soil, so much so that when canals were dug years ago, a flock of sheep would be driven over the clay lining to compact it better. There are still sheep-foot-rollers used in construction, but now with engines attached. Around the zoo, there were multiple examples of compaction. The lightest was under a tree with a bird feeder attached to it. Spilt food fell to the ground and there were always pigeons feeding there. Between walking on the soil and taking a very little away on muddy feet, a shallow hollow developed. When it rained, it filled with water. That's right, the pigeons caused compaction and puddling. Easily solved with a heavy mulch – the rougher the mulch with large chips, the better. Though, it needed replenishing every year.

The lake edges years ago, with poor grass, had a combination of compaction from visitors' feet and the wet feet of geese. The degree of compaction was such that it was impossible to dig manually. Machinery was needed to break up the surface first, and that was only possible away from tree roots. A secondary element here was the lack of organic feeding for the soil. Leaves were collected, the grass was mown and collected, and rain literally washed the surface clean as the grass was grazed so tight. The grass was poor as well as being overgrazed, so the root growth was also very poor, which affected grass growth. The soil was far too compacted. One problem led to another in a circular fashion, making it far worse.

Different animals cause different degrees of compaction. Tigers and lions walk a regular path, as already mentioned. This is impossible to stop. The only solution is to engineer out any potential damage when building the habitat and even that would need regular attention, topping up any substrate used to ameliorate the compaction, reduce or slow its formation. Among the old zoo exhibits, the wolves had almost bare soil as virtually all vegetation was destroyed by a combination of regular traffic and compaction due to regular and constant pacing This made it almost impossible to get anything new established. Tapirs were always a muddy problem once wet weather set in. The sloping bank had developed a set of terraces, giving the animals multi-level tracks to walk on following the contour, which were always muddy and bare.

The savanna was a real test when the zoo expanded in 2000. The new area for the giraffe, oryx, etc. had been a grass paddock for horses for years and was carefully kept for the savanna animals. The horses had been kept off the paddock if too wet, though the giraffes, rhinos, and zebra were out every day, all day. All animals have an in-built clock, and they all gather at the gates when it is near going-in time, as food would be waiting for them in their houses. Gate waiting areas are a recognised problem in zoos and within a short time, there was plenty of mud at the gates to the hard standings here too. When the new savanna was built, it was deliberately made with good drainage and a metre of sharp sand to allow for surface drainage too. That all worked incredibly well, as expected, until it was suggested that there should be some grass to make it seem more natural – some had been sown in pure sand, nothing really grew. Surprise, surprise. The idea was to follow nature and let grass grow in the lowest parts, where water, in nature, might collect a little. Simply dig some topsoil in was the instruction...

The horticulturalists worried that compaction would occur and suggested instead to grass the naturally draining slopes, less animal traffic, less chance of compaction, more chance for grass growth. The natural look of grass in the wet hollow won the day though and several loads of topsoil were rotavated in, graded nicely, and the grass was sown. This was done in spring. The result was incredibly verdant – a good dressing of animal manure was rotavated in too, no doubt. The giraffes, ostrich, zebras, and oryx all had access, and after a little rain, there would have been the start of compaction, but not enough rain to show it. That autumn, it really rained. There were actual floods in parts of Dublin. The compacted soil held the water so well there was a 15m wide pond there all winter. After a while, there was a request to drain the water, but the soil was too wet for any machine to go through without bogging down. The pond eventually dried up in spring. Only regular driving through with an aerator every year stops it flooding, and the grass has never done that well since. Not surprising.

Other areas within the savanna have also started to show signs of compaction on the pure sand. This is really a mixture of sand, manure, and some of the soil from the hollows. Wet feet will collect a little soil at each visit and drop it elsewhere later. This mix is the problem as it allows compaction. Annual work now with a stone burier on a small tractor keeps it looking good, but with more organic matter each year... A losing battle.

Yes, even with a metre of free draining sand, it is possible to have standing water on the surface.

If the savanna was the test, the eastern bongo *Tragelaphus eurycerus isaaci* was a major worry. When the zoo expanded in 2000, there was a woodland area on the northside which was developed for the bongo and the African lion *Panthera leo*. Developed in this case just meant fenced off. There was no thought to the trees, protection of bark or roots – or soil. The builders showed no respect either, driving over root plates at will. The lions scratched tree trunks so much (just as domestic cats do, sharpening claws), they damaged the bark of a couple of trees, including a fully mature sweet chestnut *Castanea sativa*, to the point of reducing sap flow and causing severe dieback. In the bongo habitat, all was grand at first. For a few years, there was just the odd patch of mud, but that was enough to cause some compaction and the loss of a small tree or two. Only the horticulturalists worried or considered it a problem. This was part of a woodland screen and valuable in its own right as a screen, windbreak, and habitat for wildlife. Once a patch had become muddy, there was no going back. More patches developed, and more trees died. Eventually, the patches formed half the bongo habitat. One very wet winter, there was so much mud the animals were reluctant to move, and the animal staff could not get in at all, so there was no cleaning or mucking out within the habitat, which caused issues with hefty amounts of manure.

Too thick to drink, too thin to plough – a sea of mud that was good for neither the bongos or the staff... or the tree roots.

With animal health issues and staff access issues, there were two very good handles to lever the problem into a 'something must be done' position. The horticulture team just wanted to save trees, of course, but it was much easier to push now there were other animal-related reasons. What can be done? Similar problems with other animals had been broached, and the outcome in many other zoos was to put in a hard stand – usually a stone surface that could be compacted and cleaned as need be. This often resulted in muddy conditions though – maybe only 10mm thick – a combination of dust from the stone, animal manure, rotten autumn leaves, all mixed up, and once muddy, that would hold more water again, keeping the animal's feet wet and causing more problems. What else might work?

Basically, some form of grass reinforcement was needed – a product that would spread the load and prevent compaction. There are many forms of porous paving, often a concrete grid with small holes for grass to grow in, as used for fire brigade access where a softening effect was required, or for car parks. Such a system allows rainwater to percolate into the soil. This is a more sustainable system than tarmac or concrete where rainwater must be drained away, possibly adding to a drainage system already under stress. Seen in several other zoos with various animals, but never looking good as the grass seldom grew well. Trying to lay such a system within the woodland would be very hard, not good for the trees, nor would the surface be good for the bongos' feet either. Similar systems can be found using plastic, with a grid of squares, plenty of room for grass growth, ideal for most peoples' shoes or car tyres, but very nasty for any animals, the rigid plastic edges could do real damage. Softer rubber mats were available, but potentially they could be pulled up and pulled apart, so not good for the animals – or the appearance.

Another way is to use a very heavy plastic mesh, designed for the task, laid on top of grass, often used for car-parking in fields, the load is spread enough to prevent sinking. The mesh comes in large rolls, simply roll it out flat, peg it out, let the grass grow through it. Depending on the grade of the plastic, it will take cars and even coaches. Usual recommendation is to let the grass grow enough to need mowing two or three times before the cars come in. Works very well indeed. The plastic is very hard and with sharp edges, though – no way would that be allowed with the animals, their feet would be damaged easily, quickly and badly. The mesh would take the weight of the animals easily though and prevent the feet contacting the topsoil, thus preventing the soil compaction.

Then the idea came. Could we cover that plastic mesh with a suitable substrate? If so, what substrate? A layer that would allow easy and quick drainage and allow air exchange as normal. So, the checking of products and asking of questions began. Eventually, the concept developed. Have the mesh laid to spread the weight of the animal and stop compaction, then add a thin layer of urban tree soil on top of the mesh to allow grass to grow on top. Contact the mesh manufacturers, see if they have any advice... They could not believe this use. No, no, no, the mesh goes on top of the grass. At the end of numerous emails, they were obviously happy to sell the product but had no advice after that. A totally new use. They were interested nonetheless – a possible new market. The other problem was the need to hold the mesh in place. Normally steel pins pushed into the soil would be used, but that was a non-runner because of the animals' feet. Plastic pegs? No way. What if the top came off, and there was a simple spike left? More damage potential. The edges had to be secured with plastic cable ties. Tie the edges every 200mm or so and cut off the surplus tie. Make sure the trimmed parts are facing down into the soil, or they could become a problem too.

Then came the questions, from various directions – what is urban tree soil? Years ago, the Dutch realised that the street trees were growing badly due to soil compaction under the roads and pavements and started researching ways to remedy this. A mix was created using different sizes of round stones, sometimes called 'Amsterdam soil,' more generally known as 'urban tree soil.'

Imagine a bucket full of marbles, you can compress it as much as you wish, but it will stay the same mass, with spaces between the marbles, no matter how hard it is compressed. Imagine that with round stones – the same result – it cannot be compressed further. Carefully design the ratio of stone sizes, from 150mm to 5mm, with a measured percentage of organic material and fine grades, and you can create a soil with a guaranteed space of about 20% for air or organic matter and easy water drainage too. The important part is the ratio of small particles – the silt, sand, and clay and the incorporation of organic matter. The principle works down to the smallest size. There are commercial grades of urban tree soil available down to 'tree sand,' a very fine material, with a carefully graded screened sand of one size only. That is the important part – if it is all one size, it is still like the marbles. It cannot be compressed beyond a certain point. Such tree sand can be laid under paving, compacting it with machinery to prepare for paving, ideal for paved areas with tree planting in mind – as used around the Zoorassic World.

The principle was sound. Though, how to convince the animal team? Samples of different grade round stones were brought in: 25mm, 20mm, 12mm, 10mm, 6mm, 4mm, etc. All very smooth, with no sharp angles or edges. Too large not passed as a foot could roll on them, too small not passed as it could get jammed in the hoof. After much discussion, the 10mm and 6mm were agreed on as a mix – and a mix was essential. What was needed was an air space of about 20%. As detailed earlier, the volume of air space is easily worked out by using water. Fill a five-litre container with any one size of stone, and it will need two litres of water to fill the spaces – that is 40% air space. But mix the stone size, and that volume becomes less. The mix of 6mm and 10mm sizes worked perfectly to give the magic 20% air space. All it needed then was about 20% rotted wood chips from tree surgeons to be added, to give a little extra for the grass roots. It would not be a perfect mix, but it did not need to be in this application.

For the bongos, a trial area was done first as there were quite a lot of people with little faith in the idea. An area only 8m wide and 20m long at the rear of the habitat was chosen. Work was carried out in early spring once the soil was dry enough. A long wait for a dry month…

A tractor-mounted stone burier was used; basically, a rotavator, but with the bonus that any stones would disappear in one pass, making the mesh laying easier. The actual rotavating was interesting, the compacted layer in places was a mere 100mm thick, solid, still very muddy, but underneath was dry soil – the compaction had been that effective at keeping water out. The soil structure was completely destroyed leaving the top 100mm a muddy mess. Even with waiting for dry weather, the tractor was bogged down badly in places. Eventually, it was levelled out and the mesh was laid out. The heaviest available grass reinforcing mesh was used to ensure no risk of animal feet ever pushing through it. Edges were tied together with cable ties, which were then trimmed and buried along all edges, so no trip points. No sharp plastic pieces could be left as a possible risk to the animals' feet.

A stone burier attachment rotavates the soil and leaves a level surface ready for mesh laying.

Once the mesh was in place, the urban tree soil was brought in. Here planning worked well. Order a grab lorry with a built-in lifting bucket to deliver 10-tonnes of a 50/50 mix of 10mm and 6mm round stone, readily available in quarries. Arriving at the service yard offload half, mixing the contents well so it is about 50/50 – no one was going to be able to accurately check this anyway. Then load one grab bucket of tree surgeon chippings for each five on the lorry. Again, an estimated mix was inevitable and would still work. Drive round to the bongos and load into a small dumper. Drive to the start of the mesh and tip. Easy. A little bit of spreading by hand was obviously needed. With subsequent loads, the dumper could drive over the newly laid urban tree soil. There was a slight dip from compression, but not enough to worry about, and it was to be expected. There would have been more damage if driving over the mesh on the freshly rotavated topsoil or driving on the topsoil itself.

The mesh in place, covered with an average 50mm of urban tree soil. The longer slots in the mesh are about 50mm long but only 5mm wide, so most of the stone will not pass through. Tree surgeons' chips give cheap organic matter.

All that was needed then was a little grass seed and rake it in. Step back, wait for rain and see what happens...

The first worry was whether the bongo would use the surface. New surface, loose stone, different... Within a few days, it was very obvious they preferred the surface to the wetter, muddier conditions in the remainder of the habitat. So much so that the animal staff very quickly moved their feeding points. It was more pleasant for the bongos. Grand. As well as being a lot easier for the animal staff – no mud to walk through. Everyone is happy, especially the real customers – the bongos. And tree roots were protected.

Customer satisfaction survey results. The bongo liked the new mud-free surface so much that the animal staff moved feeding points there – it was much easier for them to access.

Grass grew, but slowly. The worry was that the mix was draining too well. The wood chips should have become moist and stayed moist, and under the stones (which are just like any mulch), the topsoil would have been moist too. Watching the bongo, they have a very delicate pace to them. Their feet seem to have a gentle kick-back, just enough to disturb the stones a little. This was deemed enough to slow the grass down – plus the fact that the bongo spent a lot of their day on this new, dry – not muddy – surface. It was a small trial area to test the system, so they were using less area within the habitat. The grass did grow, but not vigorously, just enough to look green from a distance.

A successful trial. Animal staff were convinced it works, and well-timed too. Next door to the bongo habitat had been the African lion habitat, now vacant. The decision was taken to have okapi *Okapia johnstoni*, the only living relative of the giraffe, there instead. If bongo are beautiful animals, then okapi must be exquisite, graceful, forest giraffes. Unfortunately, they are heavier than bongo and potentially more damaging to the soil with compaction. They have larger, wider feet too. Despite the lions having damaged a few trees, this area was much denser woodland, great for the okapi, but a great worry for the soil and the trees. After much discussion, it was decided to use the same plastic mesh in the entire okapi habitat and use urban tree soil throughout. Animal staff wanted to avoid mud for the animals and themselves. The horticulturalists were happy to be able to preserve the trees. Hopefully.

First, what is growing in the old lion habitat? No horticultural management there for fifteen years. Lots of ivy *Hedera helix*, and other wild flora. Okapi are strange animals. They roam dense African Rainforest, browsing a very wide range of plants. In zoos, they have a reputation for being very picky over their food, for trying the wrong plant – and getting sick. A thorough trawl of the old lion habitat revealed a few plants that needed complete eradication, just in case. In addition to ivy, there was bittersweet *Solanum dulcamara* with nice red berries that birds love and spread the seeds, and cuckoo pint *Arum maculatum*, a spring-growing woodland plant that is pleasantly green early in the year, so more tempting. Both have toxic properties that were definitely a risk. The ivy was so rampant that tree surgeons had to remove it from the trees, and a mini digger with a rake attachment removed most of it from the soil surface. Two grab lorries of 10m^3 each were basically full of the removed ivy. The mini digger was also good for a little levelling and loosening – but very carefully near any tree roots. With the work being carried out in May and with much drier weather, this was a much more pleasant task than the bongo work had been. Easier to get a good result too. Indeed, topsoil likes being handled when dry.

Once the vegetation was cleared, including any surrounding areas in case the odd branch grew in, then bring in the tractors and stone buriers again. Very useful here. 150mm of 25mm round stone had been laid on top of the soil in places to stop mud near viewing windows. The ivy had completely covered it. The stone burier still found enough soil to bury the stones, but only just. After a few passes with a power rake, a beautiful surface was ready for laying the mesh down, but there was one big difference – it was not all flat. There were a couple of ha-ha's – a 1m deep ditch with a gentle slope for the okapi and a vertical side to the visitor viewing so that no glass would be needed as a barrier. The ha-ha is an old method of 'fencing' out animals, usually cattle or deer, in large estates. The aim was to allow an unobstructed view from the house out into the parkland beyond the immediate garden or parterre. There were quite a lot of rolling slopes too, including a low mound and a lot of trees to work around. One worry was what would happen if any tree was girdled by the plastic mesh, which was strong enough to stay put. Potentially, as the tree's girth increased, it would grow around the mesh, eventually forming weak points. Therefore, each tree was given a large circle free of mesh to allow for many years' growth.

The rolls of mesh were laid out, trying to get the straight sides parallel for ease of securing – every 200mm or so, they had to have a plastic cable tie attached and trimmed. Buying several thousand of these before they were needed was an essential little bit of forward-planning. Wherever there was a slope, the angle of the mesh was altered and had to be allowed for. Little triangles would need cutting out – secateurs were perfect for doing this as it was tough plastic. Very often the same triangle would be perfect for a gap elsewhere. Very little was wasted. It all took days to do. Cutting the mesh and putting the ties in place was literally a pain. Kneeling to do the task was the only way, and the mesh was sharp on the knees. After an hour, everyone had some sort of cushion – a coat, heavy paper sacks – whatever worked that moved around with them.

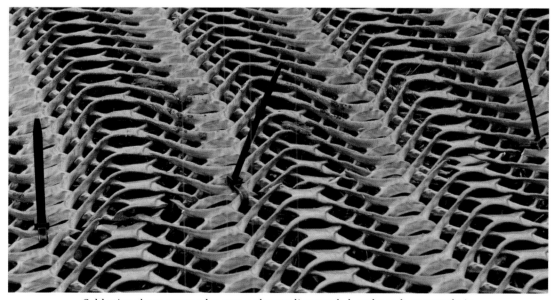

Cable ties along every edge was a slow tedious task, but the only way to do it.

One side effect of all this was the visitor perception. They looked in and saw a mass of plastic mesh, especially on the mounds. What did it look like – a ski slope? The staff could have had a nice side-line letting people try skiing on it – bring your own skis! No one could guess what was being done.

Same urban tree soil mix, though here about 50-tonnes was needed, plus tree chips.

Urban tree soil in next, again placing the urban tree soil down first, then driving on it if needed. This was a much larger area, much more driving, but the soil was drier than the trial run in the bongo next door. There was a little more marking where the dumper drove a path regularly. Nothing really significant. Simple enough to spread a little more stone. Once finished, cover with a mix of seed. This was the perfect time and weather for germination. More Rhizomatous Tall Fescue RTF was used, but with all those trees, it was much shadier, so a shade-tolerant grass seed mix was used – whichever was happiest would grow best. Hopefully. As proof that the animal team were now confident that this would work, the okapi hard standing (a small off-show area near the animal house, used at need) was also laid with the mesh and urban tree soil. There were trees here too, but no grass was expected to survive with greater animal use. Previously, to combat mud in the bongo hard standing area, a standard limestone chipping topped with a fine dust had been used. Within a few months it was already holding water and was muddy for the animals' feet and hard for the animal team to clean. It simply did not work. Seeing is believing. A thin dust layer, manure, plus regular foot traffic and waterlogging happened very quickly.

A circle kept free of mesh around each tree to allow for expansion, the hard plastic could be a problem if the roots grew around it. Filled in with urban tree soil so there would be no trip points for the animals and little risk of foot traffic in that spot either.

The grass established very well in the okapi habitat. The real proof of the effect of the mesh came during the first winter – an exceptionally wet one. The ha-ha's were soon half-filled with water – it was the natural water level in the soil in that area during heavy rain. Despite the heavy rain, the staff could still access any part of the okapi habitat or hard standing on foot or with golf buggies – no mud, no getting stuck, no foot issues for the animals. And the whole point from the horticulturalists' point of view – no compaction and subsequent root damage for the trees. Perfect.

The following summer the grass was like a meadow, flowering well, and developing a really thick sward. It looked great, the okapi loved it, but the animal team found it a lot harder to clean – daily round-ups were routine. One member of the animal staff complained that the grass was too long. The comment was, 'It's like a jungle in there.' Maybe that was stretching the definition of jungle... The answer was a question, 'Well, what is the okapi's natural habitat then?' Point made. Happy days.

A lush meadow for the okapi. Hard to believe there is urban tree soil and mesh under all of that. The mesh suppliers came in to have a look and took some pictures, but had to take our word for it.

In the same wet winter, next door in the bongo habitat, the only spot that was accessible was the little trial area of urban tree soil and mesh, so wet the mud was 200mm deep in places. Awful. It was very quickly decided to fully mesh the rest of that habitat when soil conditions were suitable. To be ready, the mesh was ordered, cable ties stockpiled. About ten trees of varying sizes had already died in this habitat, mainly due to soil compaction. If the mesh was rolled out all over, it would make tree replacement very difficult afterwards, so new trees had to go in at the same time. Four root-balled *Tilia x europaea* 'Pallida' (as used to screen the giraffe house in the savanna) at 4m tall were ordered. The work started in a dry November. That was a nice change.

All soil preparation went well, if a little sticky and much harder on the stone burier which easily gets clogged up with wet soil. Mesh was laid, a grab lorry used again. Very sneaky here – experts at that by now... It lifted the urban soil over the fence at the front of the habitat, saving much dumper driving. Some areas were not stoned first, and with the slightly wetter soil, these did compact a little. Gentle rain for the last day did not help either and after the urban tree soil was finished, there was some waterlogging in one or two spots. Manual aerating was done, but it was not easy through the spaces in the mesh. Grass seed was sown, though with less hopeful being November. With colder weather and rain, it did not germinate as well. Nor did the bongo help as their feet continually gently flicked stones around.

Even in the more wooded parts, the urban tree soil and mesh worked well, but with less light there was far less grass growth.

Spring came around, more grass seeded in by hand. Strangely – despite a dry spring, one of the wet spots still had water pooling on the surface. Very strange. It should be well dry by now. Suspicious. A little dabbling with a boot and a very gentle flow could be seen. Aahhh, is there a mains water leak here? After checking again later, there was indeed. This area had been developed in the initial build. Very poor plans, if any, for where the water mains were. A little digging revealed a joint had spread open a little. Builders were called in. The first digger bucket caught the pipe and ripped it out fully. There really was a leak then! Water everywhere. It dried out eventually, but that certainly explained why the area had been so wet for so long, even before the work had started on the mesh and urban tree soil. Once repaired, the area dried out well. No more waterlogging.

Bongos are curious animals, absolutely lovely to look at, but with hefty, gently twisted horns – unusually on both the male and female. Sometimes the males like to flex their muscles and one of the newly planted trees was getting too much attention. Each tree had been supported with a hefty stake on each side and a cross piece of timber. The male bongo had simply put his horns under the crossbar and lifted – the bar, two stakes, and the tree, all came out together. Replanted, they may lose interest… As if… It was out again a few days later. Right, how can it be protected? Impossible to erect a good, natural-looking bongo-proof barrier. With a quick look around the habitat, there was a lot of holly *Ilex aquifolium* growing as scrubby shrubs and left alone, not browsed or worried at all. Plenty more were growing behind the habitat. Cut some 2-3m long, 'plant' them next to the tree, push well into the soil, tie them securely to the stakes and the tree, fill in any gaps with more – looks like a holly copse with a tree in the middle. Wait and see how it works. And it did, a little bit of damage, more to the holly than the tree. The holly stayed green for long enough, and was easy enough to replace occasionally. How long this will need to be done is anybody's guess, though.

The use of urban tree soil had become much more normal after this. Where else can it be beneficial? The next areas were the Sumatran tigers and Asiatic lions. In both cases, but especially the tigers, it worked incredibly well, just needing topping up a little in places. Unfortunately, sometimes the topping up was done with wood chips from tree surgeons, which never works in the long run as it rots down over time, becoming finer particles and eventually, mud. A learning curve for some people but getting better slowly. The concept needed less arguing in the flamingo habitat as it had worked well elsewhere already, albeit a totally different product. Not only was the zoo team becoming used to the idea, but the builders had become used to it too.

This became very useful in another project – the Roberts House, originally for large cats, then a free-flight aviary for thirty years, and now it was to be reimagined as a reptile house with a difference – Zoorassic World. All things dinosaur – life-size models, sand pits for kids to 'dig' for fossils, and a vast amount of information on reptiles. Planting was minimal and only in the various small animal habitats, generally no more than 3m square, except the West African crocodile *Crocodylus niloticus suchus* pool.

As an iconic house with beautiful red brick walls and decorative features, it was decided to keep it fully on view. No mass planting for a change. But, with a pedestrian area outside and a Jurassic or fossil theme, it was a perfect place to grow some fossil trees – the maidenhair tree *Ginkgo biloba*. Truly a living fossil, all other plants related to it have died out. Fossils have been found dating back 270 million years. A very decorative tree too, uniquely shaped and veined leaves – hence the name. They resemble maidenhair fern *Adiantum* leaves. The autumn leaves turn a beautiful buttery yellow before they fall. Potentially a very large tree, eventually to 25m. Good soil is needed. But which soil to plant the trees into? There were to be two paved areas, for which the builders needed a surface they could compact and level before paving. Urban tree soil came to the fore again, this time as commercially available tree sand with all the grains one size – like miniature marbles, they cannot compact. The budget was put in for a half metre of tree sand over most of the paved area, giving great scope for tree roots for years. The addition of finely screened organic matter will encourage good root growth, but extra feeding will be needed in the future.

Inside planting was very different. The small habitats had tiny built-in slots for plants, some a mere 50mm wide, maybe 300mm long, and only 100mm deep – more like pots. For the benefit of the various reptiles that needed a humid environment, regular automatic misting and watering was built i. With a little tweaking, this kept the plants very happy. Some grew well, and some did not. The crocodile pool surrounds had far larger planting pockets – up to half a metre deep, wide and long, around and above a pool area. For all of these, the same issue arose, what soil to use? Tropical greenhouse habitats in zoos seen in Europe often used inert substrate for drainage and support only, with nutrients provided by a mulch. Normal potting compost has either a peat base or, if peat-free, an organic base. In time, both will rot and the compost will sink. Not desirable. Standard loam composts still have some peat in them and can become waterlogged in tropical conditions. Tree sand came in useful again. It cannot sink with time. It will drain well at all times. It will not compact. Most usefully, it was just outside the door and easily loaded into pots for carrying in to fill the relevant areas. The crocodile area took the entire horticulture team to fill the planting pockets, one person in the planter, one on the ladder passing the pot up, and three more loading and wheeling the pots in – five tonnes shifted by hand in about four hours.

All that worked well. Subsequent growth has been great. A tropical greenhouse in effect, with one plant doing especially well – the chestnut vine *Tetrastigma voinierianum*. This is a giant member of the grape family, with large leaves to 20cm or so as five obovate leaflets, attractive in their own right with brownish hairs on the underside. Potentially to 20m tall at least, this plant has been trained along the top of the crocodile's habitat wall to soften the outline nicely, with the odd side-shoot hanging down. So good, it has extended to the next door habitats too, and starting to aim for the roof girders – could be too good. *Tetrastigma* is best known as the host genus across a few species for the largest flower in the plant kingdom, the incredible parasitic – no leaves at all – corpse flower *Rafflesia arnoldii*. The single flower, arising from the vines' roots, has been recorded at just over 1m wide. The horrible smell of rotting flesh attracts flies as pollinators, hence the common name. Another giant with the same common name is the titan arum *Amorphophallus titanum*, which generates a little heat inside a tall spike, causing the scent to rise on the heated air through the dense forest, calling flies in to pollinate.

CHAPTER 15

Mulching
A Wonderful Natural Weed Control System

Mulch is a clean material applied to the surface of the soil. It should suppress weeds, reduce water loss and irrigation needs. For a public area, it should look natural and pleasant to view. A mulch can be plastic or paper sheet, but that looks awful and can blow away. Often landscapers use a woven fabric, with holes cut out for any planting and bark mulch on top, which is great until weeds start poking out of the holes as well or around the edges. No, thank you. In the early 1980s, the zoo disposed of all the organic animal waste. For mulching, several ten-tonne lorry-loads of spent mushroom compost had to be bought in every year. Great to use, easy to spread, with the bonus of a crop of mushrooms if left stacked for a few weeks before spreading. Staff watched to see if anyone was sick after the first picking... With few planted areas, this was sufficient for the needs then – mostly rose beds. It was a fairly nutrient-poor mulch as the mushroom crop had taken a lot out of it. It was also rapidly broken down, being mainly straw from horse stables, well-shredded and well-rotted, so almost gone within a year.

Seeing a large amount of manure going off-site as part of general waste disposal seemed a terrible loss of a valuable resource. By the mid-1980s, a heap had been created instead. Far more was generated within the zoo than used within the gardens. Any surplus was now collected by a mushroom company – with the caveat that it was still fresh enough to speedily rot, an essential part of their production system. The remaining semi-rotten manure was turned again to speed decomposition and was then fit to use for mulching. Never a simple heap to keep on top of. The worse problem being the elephants. There is no nice way to look at it. The produce they gave was compacted so well that it was still identifiable and un-spreadable after six months. It was never a popular component of the heap for the gardeners . While this heap gave a large volume for use, it was not perfect – often semi-rotten patches, too dry maybe, or too much straw. A large shredder would have made it a great product, but this was in the middle of the hardest financial times for the zoo, so no way to get one.

There were no other mulches available at that time. The only other bulk material on offer would have been peat, not suitable or sustainable at all. One supplier trying to sell peat was totally baffled by a request for bark mulch –

readily available in Europe but not heard of in Ireland at that time. That was to change within a few years, though, as more people started to look for it.

Then the tree surgeons started to use woodchippers to reduce the bulk of their waste. Up until then, it was a case of throwing branches onto a small lorry and using the chainsaw to rip it up to get more in. There was nowhere to store any tree chips, so it needed to be used straight off the lorry, neither as good nor as attractive as rotten material. There was a massive difference in the end product with so many factors involved. The sharpness of the chipper blades changes the size of the woodchip. The wood being chipped – hardwood is always nicer than conifers as the evergreen foliage chips badly, often with long stringy bits of material left. Summer wood chips often have a lot of leaves too, so it rots quicker. One year, several loads of wood chips from the regular pruning of street trees offered pure London plane *Platanus x acerifolia*. Pruned in summer, masses of green leaves and no wood probably older than a few years. Those big leaves chipped beautifully and with a little rotting, the loads became almost a peat-like substrate. Wonderful to use.

Bark Mulch

Bark mulch became available at last and in bulk too. This was mainly as playground material, large pieces that gave a better bounce. One supplier was found who chipped and screened his bark, selling some as play grade and keeping the rest for landscaping. At last, an alternative to manure. One other sneaky fact... At that time, the playground surfaces were maintained by the horticulture team. Within the annual budget for play grade bark, a few mulching grade loads could get quietly lost...no one would notice. Hopefully. About the same time, there was a small move towards different substrates for the animals – back to the five freedoms again. Bark was becoming a regular choice, especially for inside areas; for instance for small primates or aviaries. Better for the animals, often an enrichment too, and nicer for staff – easier to walk on and usually less cleaning as the bark created a biological layer that broke down small quantities of waste material naturally.

Care was still needed when ordering. There were options around that were cheaper but not so nice. One supplier had shredded pallets on offer. Very cheap, but if the timber had been treated with wood preservative there was a risk of a coloured leachate flowing around, staining the pathways (as seen in other zoos). Not nice. Another issue was how the bark was treated. If mechanically stripped, shredded, and graded, it became coarse play grade or fine mulch. When one playground was built, a different supplier of bark was used by the contractor. Cheaper, as the bark had been stripped with water pressure apparently, making the bark fully sodden. Stacked and shredded, graded, whatever – it stayed sodden for ages, and smelt accordingly. For weeks, the aroma wafted around the entire area. Definitely not nice.

Manure

Then the big move up to the plains. With a lot of pushing, there was an area for a stack of manure at last. Way too small, it filled in one month, but nearby room for further stacking. At the same time came a bigger change in animal bedding. The use of bark mulch became the preferred substrate in herbivore houses too. Initially, bark was the only substrate used. It took some time to convince the animal teams that woodchip from the tree surgeons would work as well in many situations – and at less than 10% of the cost. Space was found for a heap of woodchip too, but with care as to the tree material. Thorns could give foot problems. Conifers might if ingested (especially potentially toxic yew *Taxus baccata*). The volume of manure generated became greater because with woodchip for bedding, there was often a very large clean out. The desire for fresh material meant a constant supply of wood chip was needed. Any older material, now partially rotted, not wanted by the animal team was ideal as mulch for the gardens.

The manure in those days always had a proportion of wood chip through it from the bedding, and that made it easier to handle. Why not deliberately mix more in? With room to stack and allow some further decomposition, a cycle developed. Fresh manure was dumped by the animal team daily. After a month or so and a lot of heating, that stack was moved further down the bay, effectively turning it and making room for fresh material. After another month, the first heap was turned again, but this time adding about 50% rotted wood chip. This heap was then best if left for as long as possible – two to three months would be ideal. Keeping enough for a standard year's mulching would mean a heap of at least 200-300 tonnes. A supplementary heap was made for large-scale projects, especially the gorilla habitat visitor-side planting; probably another 200-300 tonnes. If more fresh material was collected, it was simply sent off-site to other gardens or a composting facility. With an increase in elephant numbers, there was a corresponding increase in 'output'... About 1,000 tonnes per year went straight off-site – far too much to handle and not the easiest to rot down. Nor was there any straw or wood chip as bedding because the elephants have a sand substrate only, not the same 'product' for mulching at the end of the day.

With wood chip or bark mulch a regular substrate for inside animal houses, there were some interesting events. When it first opened, the new gorilla house took 50m³ of wood chip – about five grab lorry-loads. With regular watering to keep the dust down, very regular donations from the gorillas, plus a heating system that kept it all tropical, the wood chip decomposed rapidly. Within two years, it went from a half-metre depth to maybe 200mm. Taken away, there was only 1.5 loads and it was the most beautiful, rotted material, looking just like peat. But why taken away? Within the zoo licencing system, there are strict guidelines on waste disposal to make sure no zoonotic (possibly passed to people) disease can be spread. All primate, carnivore, feline, or canine waste must go to a fully compliant composting facility, where the waste is held at a controlled high temperature for long enough to kill all potential pathogens.

The zebra house had a good 10m³ of bark mulch, about 300mm deep, which became dry inside, broken down by the hooves over time, it became rather dusty, so regular damping down became routine. The bark was unrecognisable by now, many small pieces and much manure – impossible to fully clean once physically broken down. So, early in the morning, when staff went in to damp down the bark (by hand), they turned the tap on – no water. Facilities were called in. Repairs were happening elsewhere, hence no water. All day, no water. Staff went home. But no one thought to turn off the tap. The leak was fixed later in the evening. By morning, the zebras were almost web-footed... Like a sponge, the whole house had soaked up a whole night of full pressure flow from the tap. The bark has to come out. Ok, but stack it separately to let it drain. It will almost certainly heat up as there is enough moisture for rapid bacterial action for the first time in months. A large heap was made well away from trees and root plates. Then a helpful soul spent half a day voluntarily levelling the heap... Two trees were completely surrounded with moist bark and bedding, about 400mm deep. When called to have a look at what they had done and asked to put their hand down into the mulch against the bark of the tree they could not. It was too hot and had been for two days. Bark heap was laboriously dug away by hand, but too late. The bark and cambium were too badly damaged by the heat. Both trees died, fully mature and healthy. What a waste.

Animal bedding is a mixed blessing. The substrate will be better for the animal but can have hefty financial aspects for disposal. Having the right substrate may reduce this, such as a coarse material that does not rot down so quickly or an acidic bark that will last longer. Much work has been done in Europe where a

information in the gardening books about some of the species available such as *L. procera* and *L. breunioides*. The information was confusing or just not detailed enough. There was confusion with blue-flowered species – only one should have been available *L. sessiliflora*, yet a second species was often offered under the name *L. caerulescens*, yet this looked very different to other *Libertia*. Ordering in new plants was always a gamble and the names of both *L. sessiliflora* and *L. caerulescens* seemed interchangeable at the nursery owners' whim. One seed packet came in from Chile. The species *Libertia tricocca* was not in the collection, so seed had been ordered. The paper packet came in with the name *Libertia chilensis* crossed out and *Libertia tricocca* written in instead. That did not prompt much faith in the source... And what grew? *Libertia chilensis*.

On the left and right L.c. Formosa Group, note the variation in petal shape, calyx colour and stem colour. In the middle L.c. Procera Group, a much larger flower.

Two things happened that focussed attention on the genus. My membership in the Irish Garden Plant Society led me to serving on some committees running the society, which led on to acting as a link with the UK group Plant Heritage, at that time known as the National Council for the Conservation of Plants and Gardens. This group concentrated on collections of plants by genera or part thereof. The idea of an official collection appealed as a way of sorting out the genus *Libertia*. Hopefully, it would be useful to all growers in the process. At about the same time, the 'internet of things' came along. The internet gave access to information about so much and easier ways of asking for help with some of the names. It was obvious that in Dublin Zoo there was neither time nor expertise to review the genus properly. Assistance was requested through the Royal Horticultural Society in England. With a heavy workload and other more important or larger genera to look at, this took time to arrange. Meanwhile, at their request, herbarium specimens were prepared. Good pictures were taken of each of the various species, and specimens were pressed and mounted. This took a long two months – noting full details of flower colour (using a standard RHS Colour Chart) and size, leaf width, and length on a standard sheet that went with each herbarium specimen. It was a lesson in patience but great to have at the end. The National Botanic Gardens, Glasnevin, Dublin, helped with relevant material and advice, as too many years had passed since such work had been part of a much younger Kew students' course.

Another year went past with specimens sent off to the RHS botanists in England and duplicates lodged at NBG Glasnevin. Then the RHS was encouraged to further the work by other reviews, and suddenly emails started – do you grow this species, with the inevitable answer, yes, but not under that name. It was surreal seeing a scan of an herbarium sheet from a botanic garden in Chile, dated maybe in the late 1800s, with a very recognisable *Libertia*, but under an unknown name for the genus, as well as the species. Watching from a distance as the RHS Registrar worked on the task, following the trail back over the years, seeing where names had been assigned incorrectly or duplicated, was intriguing and incredibly educational. A visit from the RHS taxonomist in May at peak flowering time was crucial. Though, zoo visitors did look on strangely as pictures were taken, notes compared, and measurements checked – all while kneeling down looking at a plant a mere 300mm tall with a couple of small white flowers.

There is no easy way to present this with so many name corrections – except the usual 'list everything' and give the correct name and redirect if needs be.

Libertia brunioides, see *L. cranwelliae*

Libertia caerulescens as found in trade, see *L. sessiliflora*

Libertia cranwelliae grown under the name *L. breunioides*, a name not recognised, and no one seems to know how, when or where, it originated. This became clear after ordering seed from New Zealand and when *L. cranwelliae* grew, it was identical to *L. breunioides*. The leaves are a brown colour unless in shade, the plant spreads gently but firmly with an orange-coloured stolon, readily forming a dense colony. Large (for the genus) white flowers, produced after at least two years from seed, are held at half the leaf height and the large round seed pods are held tightly shut for years. One large clump has been severely reduced by a fungal infection, possibly verticillium wilt, that slowly spread through the bulk of the plants.

Libertia cranwelliae, from New Zealand seed. So different – flowers are held low down among the foliage, very few flowers too and yellow creeping rhizomes. Rusty brown flushed foliage looks good within the savanna planting. Seed pods are held unopened for several years, yet the seed is still viable.

Libertia chilensis, the complicated one, is a very variable species, widespread in Chile, and the most typical species in cultivation. The flower size varies, as does the colour of the flower stems, the height at flowering, and leaf length. Also grown under the names *L. grandiflora* or *L. formosa* and sometimes *L. elegans* – all now technically incorrect. Imagine, almost 200 years ago, botanists around the world were discovering and naming what were new plants to them. There was little chance of accurate comparison, no easy pictures, and certainly no emails. Even worse, this also applied to the genus name. According to the rules of botanical nomenclature or classification, a name that has been used and published, often has to be maintained as it was first used, hence the many changes. Plant taxonomy often needs a careful search through published descriptions and names to determine the correct name. The confusion with *Libertia* has several elements. *L. grandiflora* stems from a description in 1856 by Philippi, who was unaware that the specific epithet '*grandiflora*' had been already used in New Zealand in 1810 for a different species (but under a different genus name *Renealmia*). The *formosa* and *elegans* epithets came from a description in 1833. The *chilensis* epithet was published first in 1810 but under the name *Strumaria chilensis. L. procera* is a name not recognised; it means tall, and it is. Again, the origin of this name is not known. You can see why there was confusion and there had to be botanical taxonomy assistance. All flowers are white, and the green leaves vary from 35cm to 75 cm depending on the group below. No stolons are produced.

With so much variability, this species has been split into three groups as below.

L. chilensis Elegans Group: A shorter plant than usual, with a more open umbellate inflorescence. Not that often found in cultivation. A slightly more refined plant, distinctive, not so hefty in growth.

L. chilensis Formosa Group: The commonest form, very widely grown, very variable in flower size, bud colour, and stem colour. By dividing a plant that is distinctive, you can develop a clone – maybe with a darker stem colour. Natural variability was evident when seed collected in recent years in Chile and grown in a botanic garden developed into plants with varying stem colour, height, and petal size. Each of six seedlings had noticeable differences.

L. chilensis Procera Group: The tallest and most vigorous grower with flowers almost twice the size of 'normal' Formosa Group. Flower stems can reach just shy of 2m tall. Regard it as *L. chilensis* on steroids. Excellent plant.

Libertia chilensis Procera Group giving a great show of bloom. Scale is everything though – this plant is about 1.75m tall growing in good conditions.

Libertia elegans, see *L. chilensis* Elegans Group.

Libertia chilensis Elegans Group, picture taken in the zoo nursery, a delicate flower, not so big and blowsy, but very fetching.

Libertia formosa, see *L. chilensis* Formosa Group.

Libertia grandiflora as found in trade commonly, see *L. chilensis*. NB: there is a true *L. grandiflora* below but very seldom seen.

Libertia grandiflora, the true species. Very seldom seen in cultivation, some at Dublin Zoo had been accessioned as *Libertia paniculata*. This is a much shorter plant and a much weaker grower, green leaves to only 25cm at most, 1-6 white flowers per panicle, far more open with the flowers hardly touching each other. This species needs to be cross-pollinated as it is self-infertile, yet the seed has been germinated. At the time of writing, no flowers as too young. Time will tell. Maybe cross-pollinated from a different species, but unlikely as in a greenhouse and no other species present. White flowers.

Libertia grandiflora, the true species. Scale is everything here, the entire plant is a mere 300mm tall at most and has not been a strong grower either.

Libertia ixioides is easily distinguished by the stiff leaves, green but brightly coloured, selected cultivars include yellow, orange, and red/brown shades.

Popular with nurseries as they look good and sell well. Several cultivars in trade, but not great zoo plants, odd colours for a naturalistic habitat planting. Although, the straight species looks grand as part of the savanna planting.

Libertia ixioides with typical brownish old leaves and as many flowers as it is willing to give.

Libertia peregrinans is commonly grown and easily recognised by vigorous stolons, quickly forming an open colony of fans of leaves. Leaves are green in the shade but a distinct copper colour in the sun. There are cultivars with slight differences in leaf colour. Same comments as for *L. ixioides* above, not a great zoo plant.

Libertia procera see *L. chilensis* Procera Group.

Libertia sessiliflora is the easiest to distinguish as it is the only blue-flowered species. The name says it all, the flowers are sessile with no stalks, forming a very tight head, too cluttered and less ornamental, though it has a nice blue spike. There are at least three different shades of blue available. The dark forms are much better. The pale form can look rather washed out, but one is different enough to be named

Libertia sessiliflora 'Caerulescens Group.' A darker form is recognised as *Libertia sessiliflora* 'Ballyrogan Blue.' They generally suffer more than other species in a really cold winter, with almost all the leaves going black, taking a couple of years to build up again. *Libertia caerulescens* has been in trade for many years and still is, but the source of the name is not known. As the name is incorrect, all are either *L. sessiliflora* or worse, a totally different genus, *Orthrosanthus laxus*. This has six petals, compared to three in *Libertia*, of a most exquisite pale blue, and the seed pods are elongated rather than generally round as in *Libertia*. *Orthrosanthus laxus* can be short-lived after flowering well for a year or two when it looks stunning. It is less hardy. The similar *Orthrosanthus multiflorus* lasts longer and has darker blue flowers in larger heads.

Libertia sessiliflora, with the eye-catching yellow stamens on a good dark blue form of the species.

Orthrosanthus laxus, a real charmer of a flower, soft blue. Unfortunately, not too hardy and short-lived – though freely produced seed germinate well for replacements.

Libertia umbellata is another easy to distinguish species, as it is the only white-flowered species with only green leaves (with an attractive slight pale sheen) that is stoloniferous, creeping gently to form a dense stand. The other stoloniferous species, *L. breunioides* and *L. peregrinans*, have stiffer, more rigid leaves, often with some colour. With that light sheen and gentle flexibility, the leaves alone will identify this species even in winter before checking the stolons. The flowers are more decorative in a genteel way, held slightly more loosely than *L. chilensis* in any form, with more room for each flower. A scanned image of an herbarium sheet dated 1871 from a Chilean botanic garden was sent to us from the RHS with a query – do you grow this? The sheet was labelled *L. ixioides* and it was thought to have been introduced to Chile from New Zealand. It was easily identifiable – very visible stolons and a delicate flower head.

Libertia umbellata, with a unique bluey sheen (compare to *Orthrosanthus* next to it) to the leaves that makes it identifiable even if no flowers. The flowers are held well clear of the foliage. Lovely plant.

Libertia x butleri is the name given to any hybrids arising from crossing *L. chilensis* and *L. ixioides*. The hybrids are variable. In trade, one is 'Amazing Grace.' These hybrids arise spontaneously. Five plants at Dublin Zoo were noticed as being a little different within the general planting of *Libertia chilensis* Formosa Group in the savanna. They had arisen from seed collected either within the zoo or by a liner nursery in their stock garden and grown on for us, within about 2,000 seedlings or so – unfortunately, no sure way of tracing back the crossing. Bearing in mind the seed collection, the cleaning, sowing, and pricking out or potting on, transport, and placement within the planting, amazingly, three of these were next to each other and two more next to each other in another area. When any breeding like this occurs with similar results, taxonomists give the cross a name to make it easier to keep track of the parentage with any cultivar name. Hence *Libertia x butleri* 'Amazing Grace,' although the cultivar was around before the hybrid epithet was decided on. The epithet was chosen to recognise the amount of work put into the revision of the genus.

Libertia x butleri, an unnamed chance seedling, with the more open floral structure of *Libertia ixioides*, but with a leaf and growth habit more like *Libertia chilensis* Formosa Group.

Libertia micrantha, not included in the review as they all died...picture taken in the zoo nursery, the pots are only 7cm square, the flowers were only a few mm wide. This species grows in shady woodland, often on mossy logs and is only up to 200mm tall. It was never going to be a useful zoo plant, but nice to have seen it. From New Zealand seed.

Libertia does have one drawback – the masses of flowers leave masses of flower stems, rigid and untidy looking, and if not deadheaded, the seeds will scatter and pop up in vast sheets if in a good environment. Deadheading 30 stems per plant on maybe 1500 plants becomes a major drain on the gardening team at a busy time of year, midsummer. To do it properly, use secateurs and cut as low down as possible. Otherwise, the cut stem stays there, rigid and now sharply pointed, to catch unsuspecting knuckles pushing in next year when deadheading again. A harsh winter, such as 2009/10 or 2010/11, where temperatures went down to -15°C for ten days or more, can severely damage many *Libertia* leaves too. Each species varied in damage. *L. sessiliflora* was completely blackened to ground level. Among the *L. chilensis* groups, *L.c.* Procera Group was most damaged. Almost all leaves browned to within 250mm of the centre of the plant. Interesting that the other two groups were not as badly affected, proof of the different source – possibly a higher altitude? Cleaning off the dead leaves of so many plants was a real pain for the team. Days were spent doing the same task.

The amount of time spent deadheading and cleaning easily convinced the team that there were too many *Libertia*. In fact, a good few disappeared after the two winters damage and were never missed. The decision was taken to slowly remove more, they were too thickly planted, necessary at first, but not needed after a few years and the individual plants look better with room to form a natural mounded shape and not be pushed about by neighbours or, often far worse – a fence line. What to replace with?? *Moraea huttonii* as detailed above fitted perfectly, a different splash of bright colour for a while and far less work.

Digging up the old clumps was a real pain– very heavy going and all by hand, of course. The large old clumps of *Libertia*, now five or more years old, were only fibrous-rooted, but that clump of roots could be half a metre across and deep – size, weight, and awkwardness against brain, brawn, and tenacity. Spades broke, forks bent, backs were sore. Eventually, the trick to lifting them was by using a crowbar and a timber block as a fulcrum, knock plenty of soil off to lessen the weight as it lifted. Even then, the team would not want to lift too many at any one time. Then a request from some of the animal staff for 'something green' to enhance the look of an indoor animal house, with no natural light and damage was also expected, so any plants used would die anyway. There's a thought – a few old *Libertia* will do, and tough enough to last a while. This not only worked, but it became an almost regular request which was grand if the team was not too busy. It also meant the lifting of the old plants was spread out over many months; much easier to fit in and easier on the team. This routine has become so regular that 'spare' *Libertia* have been planted in off-show areas for future use when there are no more to be lifted in the general planting.

The most unusual request was when an oryx was born, and the animal team realised that it needed a little bit of cover. 'Could we please have...' and a few clumps of *Libertia* went in. With the natural arching shape of the clump, there was a perfect place or three for the young to hide. Excellent. Back to the five freedoms of animal keeping again. The only odd thing was that the parents started eating the *Libertia*. But surely nothing eats *Libertia*, does it? Well, consider the oryx and its natural near-desert habitat. Maybe, just maybe, it is more used to really tough food at times, even if it does have crystals of calcium oxalate in the leaves. Good job that it is not poisonous though...

Other animals that enjoyed the clumps were the slender-tailed meerkat *Suricata suricatta*. These energetic animals are always searching for food – always a good place to throw snails if found around the garden as they love them, but with constant digging in their sandy zoo habitat, no plants can survive for long unless protected. 'Plant' a couple of old *Libertia* clumps, let them search for any insects for a few days, then scatter a handful of mealworms into the clump and watch them go. Maybe not a pleasant way to go for the plants or the mealworms, but, hey ho, five freedoms again – enrichment to allow natural behaviour. Taking pictures of this was the most entertaining task – especially for any visitors watching. While kneeling to take a steady shot of several fast-moving small animals, a few of the meerkats wanted to check the camera settings. With one balanced on a shoulder, another on an arm, two more climbing up onto the photographers' legs, and a queue waiting for a free spot… Unique.

Excellent enrichment for the meerkats. The plant roots were great hiding places for mealworms. They resisted digging for a good while, hours of delight – for the meerkats that is.

CHAPTER 17

Weed Control
A Never Ending and Essential Task

Times change, fortunately, sometimes for the better. Weed control used to mean either hand hoeing or pulling up, but that is so labour intensive. Alright for small areas, with small weeds, but not for large areas in a public garden, unless there is plenty of staff and time. There were several weed issues years ago, each a different headache. Scutch grass *Agropyron* (*Elymus*) *repens* was well established through a lot of the *Cotoneaster* around the zoo. The treatment for years was simply pulling hard to remove the most visible growth. It certainly looked better – for a little while, but it was never under control. A glyphosphate applicator developed for roguing common wild oats *Avena fatua* was sourced. This consisted of a sponge on a glove, all you had to do was squeeze gently and the glyphosphate was applied to the leaves. All were killed within two years and much less work – plus it was only spot treated, no spraying.

Bindweed *Calystegia* (*Convolvulus*) *sepium* was a worse problem in a few areas and very noticeable when in flower. Dabbing each leaf with glyphosphate eventually worked. It had been very slow, awkward, and fiddly, but very effective – until the CDA or Controlled Droplet Application system came in (see below). Path spraying for annual weeds was carried out for years with a standard knapsack sprayer. This meant messy mixing of weedkillers and 20 litres of water on the gardener's back. All mixing had to be done in the nursery, so there was a walk of maybe a few hundred metres before spraying could even start. Don't even think about the boundary fence line... Pushing through overgrown bushes and stepping over or getting under angled fence supports – all with 20kg in the knapsack. Woeful.

Then a new application system was developed and put on the market, using a more concentrated weedkiller, but thus needing much smaller doses and more importantly, lighter to carry as no water was needed. No mixing either. Wonderful. Though it was a little more expensive, the time saved more than balanced out the money spent. This system has since been used for all weedkilling needs in the zoo. The CDA Controlled Droplet Application equipment was usually only for spot treatment and not general spraying. It works extremely well. Eventually, only contact glyphosphate was used and no residual weedkiller. Even better, less residue. A simple gravity-fed lance holding 750ml of weedkiller and a battery-powered rotating head with the right setting gave a 150mm circle of spray that dropped to ground level immediately. A very small amount would go a long way, the good gardening policy of never letting a weed set seed was followed religiously. And it certainly worked, reducing work for future years. The adage 'one year of seeds is seven years weeds' is certainly true. There were very few gardens using the CDA system, though. When training became compulsory and a trainer was found, the first thing he had to do was come to the horticulture team to see how it was done having never actually used a CDA unit. Even then, the team had to help with the calibration checks. The spot treatment was the only system used, so the calibration was never needed. It purely gave you a speed of walking to stick to for a definite application rate – something not needed as only the weeds were sprayed, never the bare soil.

This was a remarkable change compared to a knapsack full of water – a handheld lance weighing about a kilo was all that was needed with a small ready-to-use pack of weedkiller to plug in. If there were not many weeds, that pack could last two hours or more of walking. The principle of only spot weeding carried on, spraying only where the weed was, no more. This kept the amount of chemicals used as low as possible but may have meant a second or third check would be needed in some areas – always a better option than overuse, though. A sustainable and responsible method of using weedkiller.

As an example, one small area just planted up in early spring had almost no weeds after mulching – until maybe 250 creeping thistles *Cirsium arvense* grew from every tiny broken root throughout the soil – always a risk with topsoil bought in. While still only 50mm high, a simple drop of glyphosphate weedkiller on each plant was enough – this was easy with the CDA system, just depress the trigger for a second but have no spin on the applicator, a drip would appear and drop. No bending down, no digging out roots and making a mess of the mulch. Two weeks later, do it again with only fifty to do. Again three weeks later, but having to really search for the odd one left. Spot application only, very small amount used, no blanket spraying. Perfect.

The bindweed had been a problem for many years. No surprise there. One hedge was so thickly covered with it that a general spray over almost the whole hedge hardly hurt the hedge at all. The intention was to remove it anyway, so no risk either way. Other areas needed more care and early morning work before visitors arrived. The CDA system worked well. If the applicator head was still wet, it could be tapped on the leaves and deposit enough to work. New growth in early summer could be easily targeted at ground level or maybe using a boot to push the bindweed growth down before spraying it. After a few years, it was almost totally gone. By that stage, it was no longer a worry though, it was controllable.

Odd weeds appeared in odd places sometimes. Buckwheat in old bedding was mentioned above in mulching, as was cannabis. Both were left over from scatter feeding. Other odd plants would sometimes appear from the same source, but none did as well as those two. The mulch could also give a weed problem. The source of the mulch is the issue. One year, with no area to stack the wood chips to rot for a while after delivery, a load or three was used directly from the tree surgeons. The next spring, there was a veritable nursery of Lawsons cypress seedlings popping up. Sure enough, that's what had been cut down, chipped, and delivered. It just reinforced the need for an area to let the material rot first, usually decomposing with enough heat to kill off seeds, or at least most of them. The oddest was still the *Impatiens* in the whale skull – what were the odds of that?

Remember, a weed is a plant in the wrong place... One year a new 'weed' was seen and, fortunately, not just sprayed but asked about. There were several nice about-to-flower specimens of broad-leaved helleborine *Epipactis helleborine*. It is such a lovely find and once noticed, more were seen around too. It is widespread around Ireland, so it was no great surprise that it cropped up. What was more interesting was the areas it appeared in. As new roads were built around the zoo, the builders used the cheapest stone aggregate available locally. This is usually crushed limestone. At the path edges, there is always a little negotiation between the stone and topsoil for planting. That edge or anywhere that the stone was still too near the surface, was the favoured spot for the helleborines to pop up. Within some bought-in plants, there were occasional 'weeds' of the common spotted orchid *Dactylorhiza fuchsii* subsp. *fuchsii*, again a very common orchid but with very many forms and variations. Wonderful to have. Besides, nobody would complain about a weedy spot if it was an orchid, would they?

Noticing unusual weeds is a valuable skill. Seeing anything that looks different can often be well worth the time and trouble, especially in a zoo and a public garden. One such case was a rather odd buttercup. It looked peculiar. Pulled up anyway, potted and grown on – purely out of interest, it turned out when it flowered to be the cursed or celery-leaved buttercup *Ranunculus sceleratus*. All buttercups are poisonous with the chemical protoanemonin, an acrid oil. This is not present in fresh material, but is produced by the action of an enzyme, ranunculin, which only happens when the plant is bruised or eaten. The ranunculin is very bitter; hence animals do not eat buttercups or their relatives. The cursed buttercup has up to 2.5% ranunculin, the most of any species. Crushed or damaged leaves will cause nasty blisters on human skin. Apparently, many years ago beggars used this deliberately to induce a more sympathetic response from people. Where did the plant come from? An annual, probably brought in among topsoil or even on machinery wheels or tracks. Well worth spotting and removing the first one, though.

Sometimes plants can be too successful. Among the numerous species planted throughout the zoo, several have already become a nuisance. Already listed above is the giant Japanese butterbur Petasites japonicus var. giganteus. When originally given to the zoo (no one would dare sell it, would they?), it had the caveat that you could plant it but not stop it. That is certainly true. For that reason alone, it is not planted in too many places, most of which have limited potential – perhaps surrounded by concrete or on an island. Where it is near a path, the new 50mm thick tarmac did not stop the flower buds pushing up, let alone the leaves. A new road surface went down, with a deep concrete edge along that section, the facilities department were ever hopeful…

A real thug is the Himalayan honeysuckle *Leycesteria formosa*. It was planted in bulk only a decade ago, but with very tasty berries, the birds have done what they are meant to do… Seedlings have sprung up in lots of spots, some good, some not so good. Each berry has lots of small seeds, which means they get everywhere. Slow to get going, a two-year seedling may be no more than 200mm high. Do not be fooled – they are just getting their toes well down. Vigorous growth takes off soon enough. Once established, they get into the 2m growth per year game, and then they definitely make an impression. On balance, they are good – popping up in the most unlikely places. Brilliant. Nothing seems to eat it. Successfully seeded into various primate habitats, rhinos, elephants… The only problem may be stopping it. Nuisance value to weed out, but easy if small enough. If big, cut down and put weedkiller on the new growth.

One of the most colourful and characterful plants around the zoo is the giant viper's bugloss *Echium pininana*, the largest number being around Society House. This wonderful monster of a plant that starts as a seed germinates in spring (better if a little frost stimulates it). It grows large, coarsely hairy leaves to a half-metre long and 200mm wide, on a stem 75mm thick that reaches 1m in the first year. Good grower and worth it for that alone. But, the second year, starting early in spring, growth takes off like mad until there is a 4m tall tower. The flowers are blue and in the hundreds over the entire summer – but best in May when it is all still nice and fresh with green leaves. Each flower has two or three seeds that fall to the ground when ripe to germinate next spring. Do the maths, how many hundreds of flowers, times two or three. Then imagine 75 plants that were in one area… At a conservative estimate, let us say 10,000 seeds. If still attached to the rough sepals, some will blow and roll around until they find a place to stay and grow. In some places, the seedlings ended up like a lawn – no mulch visible. Some were wanted for the following year's display. In time, they will thin themselves out. If left alone for a month, some will grow faster and crowd out their neighbours. Not so much left then to weed out, instead edit out the extras. Leave three or five together – they will grow apart in the most charming natural fashion. Wonderful. But, oh, what a potential weed if not watched. A really hard winter will kill them all. *Echium pininana* is endangered in its native habitat on La Palma in the Canary Islands, where it pops up in woodland clearings, but is an invasive alien in some places around the world now. Bees love it. Gardeners love it. Wrens have even been found nesting within the rough pendulous leaves. Let us put up with it a little longer please.

Echium pininana performing well. Several seedlings left to grow had grown away from each other, giving a very natural look.

Mulching was the first and primary weed control system and was so effective. Chemical weedkillers were only used as a secondary treatment and much less needed after a really good mulch.

CHAPTER 18

Fauna and Flora
Uninvited, Often Welcome, But Not Always

Visitors come to the zoo to see the exotic wild animals kept there and, hopefully, return home enthused with the idea of conservation and care of the environment. Any garden of thirty hectares, especially with water and a diverse planting list, is bound to have some native wildlife enjoying the habitat too. Already mentioned are the little grebes and other waterfowl in the moat around the gorilla habitat. Over many years, there were many odd or unusual meetings with insects, animals, or plants. Insect enquiries often came to the horticulture team. One phone call was from a very worried gentleman. He had found a large beetle – this was way before digital photos, and mobile phones were invented. When it was suggested that they photocopy it, there was disbelief, but they did. Sure enough, it was a male cockchafer *Melolontha melolontha*. Then the question was, 'How do you know it's male?' The antennae are large and fluffy, so it is male. Put it outside, and it will fly away. 'It can fly!' was the astonished response. A very common insect, people just do not look.

The animal team were watching the gardeners replace some of the indoor plants many years ago when something jumped out and ran. Watch the jumping spiders was the comment. Spiders? That big and jumping? No, actually house crickets *Acheta domesticus*, common enough in warm buildings with plenty of food left around too. Plenty of German cockroaches *Blattella germanica* too, a real pest and hard to control, only to be expected in warm animal houses or tropical greenhouses. Common wasp *Vespula vulgaris*, were the commonest insect that had to be dealt with. If nesting somewhere out of the way, they could be left alone. Highly beneficial in summer, eating flies to feed to their young. Nests were most often found when cutting hedges – by which time the nest could be very large indeed. Visitors were more at risk if they were eating ice cream or, worse, candy floss. One year there were so many wasps, the coffee shop had to stop serving candy floss and even closed for a while. The wasps knew the time of day and only became a nuisance after 10.30am when the machine was turned on. Another time the team was asked to check the African spurred tortoise *Centrochelys sulcata* as they were being bothered by wasps. This did not make sense at all. This was in spring, with no great wasp numbers around. It was checked out, and sure enough, there were a few wasps buzzing around the eyes of the tortoise, but why? Looking around there were another fifteen or more dead wasps along the windowsill. All were queens, obvious from their larger size. No worker wasps about so early in the year. The only answer was that the queen wasps had all hibernated in the thatch-effect roof of the building, came out into the warmth inside and immediately started to look for a drink after a long winter asleep. The only water was a small drinker operated by the tortoise at need. The only other moisture was the tears on the tortoise. Hence, they were chasing them, landing on their faces and drinking their tears. Not sure it did either of them any good though – the wasps or the tortoises.

All gardeners see odd insects while they work. It makes the day more interesting. Instead of 'cutting the hedge,' it is a mini biodiversity check. Within the Family Farm meadow a few wild roses had been planted, common dog rose *Rosa canina*. The flowers looked grand on top of the hedge. Care was needed though to keep the thorny shoots in the hedge or they attacked anyone passing by. Within a couple of years, there was a regular specimen or three of bedeguar gall or robins pincushion *Diplolepis rosae*. This is caused by a tiny gall wasp laying several eggs in a shoot. The gall, a spiky feathery-looking growth maybe 50mm across, is a chemically distorted bud. The larvae are protected within the structure and almost always females, very few males are ever seen. Leave well alone – it is no harm to the rose at all. It is never seen on cultivated roses either so no risk. A real curiosity and good to see, and useful for educational tours too.

The year after the sea lion habitat was finished, the 'stream' connecting the lakes seemed to have a white oily coating – a little worrying when the flamingos were a short distance downstream. What was causing it, though? There were no oil leaks nearby, and it being white was very strange. A closer inspection found lots of tiny white flakes on the surface of the water. Using the always handy x10 lens, the cast-off skins of an aphid were identified. But which aphid, from where, why so many?? Leaves of *Darmera* and *Gunnera* acted as great traps, similar to upside-down umbrellas, for this rain from above – so look up and ahhh... A Lombardy poplar *Populus nigra* 'Italica' was infested with them – but they could not be seen. Most unusual. A closer look at some fallen leaves revealed poplar gall aphid *Pemphigus spyrothecae*, which have a most unusual hiding place. In spring, overwintered females start a colony on the petiole of the poplar leaf, which deforms into a flattened, twisted shape, hollow in the middle. Untwist carefully, just enough to see in, and there are the aphids. Safe inside, they even have 'soldier' aphids guarding the entrance to the hollow. The aphids produce a waxy secretion as added protection, like woolly aphids, of which there are many different species. The 'polluted' stream surface was simply the detritus of cast skins and waxy particles. No harm. Anyway, how would you treat them?

Poplar gall aphid *Pemphigus spyrothecae*. The deformed twisted petioles keep the aphids safe from predators.

The first autumn after the sea lion and flamingo stream was made, early mornings would often see a kingfisher *Alcedo atthis* perched on a rock near the bridge – but gone in an instant, a flash of azure blue. Maybe it was the same one that had a regular fishing spot that winter at the hippo pool. Wonderful to see such an exotic-looking native. A very infrequent visitor – or is it? Very hard to spot if perched, only in flight is it easy to see.

Rabbits were seldom a problem as the zoo is fully fenced to keep out foxes, but one year at least one rabbit got in. Sure enough, as per their reputation, in no time, there were loads. Initially, they caused quite a lot of damage in the savanna and, as always, targeted the choicer plants. For instance, all *Gladiolus tristis* disappeared for a while. As only plants were affected, and no one else but the horticulturalists could see the damage, there was little concern at first. They became more adventurous with more numbers and, suddenly, there were animal issues too; contaminating hay, for instance. The real push came though when they started to disturb the elephants. Imagine it – at night, a nice, tasty carrot was just out of reach of the rabbit. The solution was for the rabbit to simply dig under this rock shaped like a sleeping elephant to get it... Control soon started then. The end of the rabbits was when a pair of foxes took up residence for a while – biological control at its best.

Badgers were living within the savanna and surrounds. They spread down to the lower lake area too. Never a real problem. Although when a large hole appeared within the main *Libertia* collection area, it was a worry. Fortunately, the badgers moved off again. The main evidence of their night-time feeding was the sheer number of scrapes through the mulch, getting at worms or grubs, no doubt, but there was no real damage. It proved the value of the mulch as a real benefit to biodiversity again. Occasionally, a wasp nest would be found and dug out for the larvae. That helped keep the number of nests down but caused a problem if anyone went near the area. Any adult wasp still there was homeless, hungry, and very unhappy – ready to sting anyone or everyone!

One morning a muddy groove was found on one of the lake islands, opposite the flamingos. Now, what would make a muddy groove on the bank?? Birds maybe, but no footprints... Maybe, just maybe an otter? They make such paths for rapid entry into the water. And indeed it was. A lovely well-fed large male was eventually caught and relocated – all obviously very officially done for such a wonderful species. Great to see.

Not so welcome was a mink, noticed next to the orangutan habitat one day. The most peculiar part of this was the reaction of the orangutans. They picked up sticks and stones and threw them with gusto at the mink. Possibly recognising it as a predator.

Although the general landscaping with a very mixed palette of plants encouraged plenty of insects, there were always some areas or even specific plants that were better. Blackberries *Rubus fruticosus* are one of the very best insect plants. Well-known by everyone too, but an awful thug within a garden and very well-armed to resist pruning. Other species can be equally as good for insects but with more interest to a gardener. The incredibly small-flowered *Rubus ichangensis*, with long panicles that seem to flower for ages, is always a hit with bees and hoverflies. Everyone mentions *Buddleia* for butterflies but equally as good for them and a far better plant, is *Escallonia bifida*, an evergreen shrub from Brazil and Uruguay. The white flowers are closely packed, giving a solid surface to land on. Better yet, flowers appear in late August and September, when there are loads of butterflies around. Not completely hardy, though, to -5°C, maybe more if in a sheltered spot. Wonderful.

The lakes were always a point of interest – a fantastic asset, with views across the water giving visitors a feeling of space as they ramble around. Water bodies are constantly changing, depending on wind and weather, sunny or dull. The edges, in particular, are potentially the biggest area for change – vegetation tries to reclaim the water, erosion tries to reclaim the land. Dynamic indeed. Excessive geese numbers had benefitted the erosion for many years. Now the plants had the upper hand, especially in some areas. Better for wildlife too. Cleaning up the lake, especially putting in the new sewer system, really helped. Far fewer nutrients are going into the lake.

Escallonia bifida and red admiral butterfly *Vanessa atalanta.*

One summer, there was a more than normal amount of duckweed *Lemna minor* – the same plant that took over the bear moat years ago. It was first noticed in the moat around the orangutan's original island – a warmer, wind-sheltered area. Each plant is made up of several floating 'leaves,' each only a few mm across, each with a root. They break up if more than three or four leaves and grow more again. After a few days, there was noticeably more. Will that be a problem? Another week and much more. Hmmm…. Then a change of wind, and the large mats of duckweed left the shelter of the orangutan moat, broke up and took off all over the lake, hardly to be seen, but still there if looked for. Another month and there were solid corners, having multiplied in areas – twofold every week, or so it seemed. Then the wind changed, and it all blew down to one end, half the lower lake was covered. Then there was no wind, and it spread back. Slowly, the entire surface was covered with no water anywhere to be seen. A verdant lawn of duckweed. Incredible. The lake area is approximately 15,000m². If one duckweed plant is about 5mm x 8mm that means a very minimum number of individual leaves of about 25,000 per m², so for the lake say about 375 million. And all with the potential to propagate and increase. This became an issue as some of the primates on an island had been used to a grass habitat. Sure enough, one of them took a step too far and fell in. Rescued and dried off. No harm done, but then the question was, what can be done? Trying to clear around individual islands worked for about an hour. Just swishing the plants around disturbed them, but they were pushed back. Probably more by the surface tension of the water acting on the floating plants than anything else. Pumping out the lake water but aiming for the surface reduced it a little but did not keep up with it. Eventually, autumn came, and with frost, all the duckweed disappeared. It overwinters as dormant buds called turions. These sank to the lake bottom, rising the following year. The next year saw only the normal amount. Fortunately.

The lake surface fully covered. Too thin to plough, too thick to drink. A remarkable sight that had most visitors wondering what had taken over.

to common clover *Trifolium*, with a sample of roots to show the nitrogenous nodules caused by the bacteria.

The education team always responded in a different way to a horticulturalist, amazed at points a gardener takes for granted. For instance, the giant Himalayan lily *Cardiocrinum giganteum* planted in the Kaziranga Forest Trail. The name is explained easily – originally from the Greek words *kardia* for heart, as the leaves are heart-shaped and *krinon* for lily, as the flowers are very obviously from that family. Then add on the botany – flowering height is one year's growth, seven years or so from seed to flowering size at 2-3m, after which it dies, but daughter bulbs carry on. Seeds are dispersed in the wind. Too much so in New Zealand, where it has jumped the garden fence and is an introduced alien plant needing control. It is difficult for any European gardener to imagine weeding out *Cardiocrinum*. All are interesting points to weave into a lesson while walking around.

Chemical Warfare – Edible, Non-palatable or Poisonous

Animals or insects want food and plants do not want to be eaten (unless in fruit). Think of it as a constant state of war between plants and animals or insects. This has caused some really remarkable changes in the structure of plants (see below on physical defence), but even more in the properties of various chemical toxins present in some plants. Many of these toxins are bitter, may have a taste like pepper, or in some way make eating the plant unpleasant, even to the point of being fatally poisonous. The toxins may be in a specific part of the plant or throughout the entire structure. Once the insect or animal has found it unpleasant, obviously, less damage will be caused.

The scent may be enough to deter potential insect or animal damage, either triggering the memory of an unpleasant meal or being unpleasant enough in itself. Plant choice within or near animal habitats had to ensure that there were no acutely poisonous plants within reach of the animals. Back to the difference between palatable, non-palatable, poisonous, and deadly poisonous again. Some insects can not only take the poison without harm, but they can also use it to their advantage to prevent predators from eating them.

These toxins can be incredibly valuable medicinally. Sometimes they have been discovered by watching animals in the wild collecting particular plants when sick. These plants are not part of the animal's usual diet and are eaten only when needed. There are many examples of this – with the lovely name, zoopharmacognosy. Studies have shown many animals use plants as medicine, though it must pass the test of being not part of their regular diet, not of nutritional benefit, maybe only eaten at certain times of the year or maybe eaten by other animals with the same checks. The majority of studies have been with primates. One well-known example would be chimpanzees eating leaves of Vernonia amygdalina, only when unwell and very occasionally, usually to reduce the parasites in their gut. This could be talked about while standing next to garden plants of a similar-looking related species of Vernonia, several of which were planted around the zoo. Other remedies include very rough leaves rolled into a ball and swallowed whole, mechanical rather than chemical. Internal parasites can be physically removed during digestion. Chimpanzees, bonobos, and gorillas have all been seen eating Aframomum seeds, a ginger relative that has shown anti-microbial action when tested. It is very interesting that three different primate species have been observed eating the same plant for medicinal purposes. Again, there are many related plants around the zoo. It must be noted that only a particular part of a plant is sought out at times – the seeds or the sap. Remember how the gorillas and mangabeys targeted the catkins of the Salix or the end of the Carex flower stems. They are well able to think it through.

Vernonia arkansana, native to the North American prairie, has very rough leaves similar to African species.

Talking about this is easy, but it becomes a lot more interesting for educational tours if there is a plant to look at that actually has these toxins. Around the zoo are many plants with a known drug element, some very common too. One of the original sources of aspirin was willow *Salix* species, used for many years under the old Doctrine of Signatures principle – the leaves trembled in the breeze, so it must be good for patients trembling with a fever. It worked too – to some extent anyway. Each species has different levels of drug though. The most is found in purple willow *Salix purpurea*, with such high levels of salicylic acid that the leaves are unpleasantly bitter, so are not eaten, as found with the gorillas and orangutans. Most interesting though, is the fact that the catkins are eaten eagerly, so there is probably no bitter salicylic acid in them.

The gorillas soon learned that *Salix purpurea* was too bitter to eat.

The following plants have been found to be not palatable to various animals. It does not necessarily follow that they are not palatable to every animal, but all have a story to tell. Some have been discussed above in relevant animal habitat notes.

From South America, Ivy of Uruguay *Cissus striata* is in Vitaceae, the grape family, not a poisonous genus. They were planted in the narrow stand-off area of a small exhibit for Bolivian squirrel monkeys *Saimiri boliviensis boliviensis*, which had a reputation for eating/playing with/destroying vegetation. They immediately came down, took one sniff at these new plants and retreated quickly. The plant was never touched by any of the squirrel monkeys. It became a dense climber covering the steel framework and mesh, giving great cover. Now, check-up *Cissus* and there is actually one species *C. quadrangularis*, that is used in various remedies, including ayurvedic medicine. It is widespread in Africa and Asia. If a related species has such chemical resources, maybe *C. striata* has some too, but has not yet been discovered by scientists.

Winter's bark *Drimys winteri* is certainly full of toxins, warburganal and polygodial. Both of these have a very peppery taste, most likely to deter browsing from animals or insects. This was not used in any animal habitat as too slow to establish and only available as small plants usually. Such an interesting plant from a very old family with fossils dating back about 125 million years and an original distribution around the southern hemisphere. *Drimys* is Greek for acrid, an apt description for the bark. The specific epithet relates to Captain Winter, who sailed around the world with Francis Drake 1577-1580. Rounding Cape Horn and with a sick crew, he landed to seek supplies. The bark of *Drimys* (which is green inside), boiled in a stew no doubt, was found to be an excellent remedy for scurvy, an awful disease caused by insufficient Vitamin C. This was centuries before Vitamin C was actually described. Now, you do not need to strip the bark to get a flavour of the peppery taste. A very small amount of leaf chewed gently will work very well, but it takes a minute or two to work. It may be best if there are some sweets to pass around to anyone complaining when sampled so as to remove the sometimes uncomfortable hot taste. A great way to make a visit memorable.

Drimys winteri in full flower is quite a spectacle. The large evergreen leaves still look good in winter.

Escallonia resinosa, planted in the sea lion cove, has dead leaves that smell of curry. It is a feature noticed most often in herbaria (dried pressed plant collections) as several *Escallonia* species have the same aroma from the dead leaves. An interesting topic. Scent varies as insect repellent in leaves or as insect attractant in flowers. Not always a pleasant scent if fly pollinated, a rotten meat smell. This particular *Escallonia* originates high in the Andes. It is an important timber crop with very hard wood, even though it only grows to a height of 10m at most.

Spurge *Euphorbia*, a common garden plant, with several species in various places around the zoo from a very varied genus of about 2,000 species, ranging from annual weeds to large trees. The milky-coloured latex-like sap has lots of potential toxins causing digestive problems or skin irritation. The seed dispersal can be explosive. Each seed pod splitting and violently throwing the seeds out. *Euphorbiaceae* is unique in having all three forms of photosynthesis in different species, though they are a bit complicated to explain as part of an educational tour. CAM Crassulacean Acid Metabolism photosynthesis evolved in dry climates. The stomata (breathing pores) close in sunshine but open at night to take in carbon dioxide, which is stored for use the next day. First discovered in the genus *Crassula*, hence the name. CAM is found in *Bromeliads* for the same reason and in some water plants as carbon dioxide is more readily available at night when dissolved in water. There are different forms of photosynthesis. C3 is the most commonly found in over 90% of all plants but is dependent on sufficient soil moisture, moderate temperature, and sunlight – no extremes. C4 developed later and is a much more efficient method, using only a third of the water needed for C3 photosynthesis, tolerant of higher temperatures and stronger sunlight. Most grasses are C4, thriving in drier habitats than C3 plants could. Think savanna grasses and scattered trees.

Euphorbia mellifera can grow to 3m or more with no severe winters. Here they are a highlight in a diverse screen along the Gorilla Rainforest.

Euphorbia rigida and *Crassula sarcocaulis*, both plants that use CAM photosynthesis. Here, growing along the edge of the savanna trail.

St. John's wort *Hypericum* is a common garden plant with many good species to choose from. The bright yellow flowers are very attractive to pollinating insects. Various potentially toxic compounds are found in different species. Why St. John's wort? Well, wort comes from Old English *wyrt* and Old German *wurtiz*, both simply meaning plant. The Feast of St. John replaced the older midsummer festival Walpurgisnacht (Saint Walpurga's night) when *Hypericum* flowers were used to ward off evil.

Flag iris *Iris pseudoacorus* has calcium oxalate crystals in the leaves, like many plants in the *Iris* family. They make it unpalatable but not poisonous. This makes it perfect for lake-edge planting, as various waterfowl will not eat it. *Iris* was named by Linnaeus for the Greek goddess of the rainbow, Iris, as there are so many colours within the flowers of this genus. The various flowers within the family have many interesting pollination mechanisms, often linked to a particular fly, bee, or even sunbird.

Himalayan honeysuckle *Leycesteria formosa* has proved to be a very useful plant, as no animal seems to eat the foliage. As with all plants, there is a cocktail of compounds within it. But all are found in everyday edible plants. Therefore, it is a mystery why primates, rhinos, or elephants do not eat it – at least not in Dublin Zoo's experience. However, it is an invasive alien plant once outside its natural range in the Himalayas. It can become overabundant in woods. It is noted as not being eaten by deer – one reason for its invasive nature, in deep shade and in competition. It has the great advantage of fleshy berries which are readily eaten and spread by birds, including pheasants. This plant was originally encouraged by gamekeepers for precisely that reason. Pheasant berry is one common name. It is easily grown from hardwood cuttings in winter – just push 300mm shoots into the moist ground in early winter.

Leycesteria formosa flowers are highly ornamental, the red bracts remain after flowering.

Not even the elephants tried to eat *Leycesteria*, here grown in a raised bed and hanging well down within reach.

Though grown mainly for the leaves, *Cynara* flowers are spectacular too.

One of the most unusual requests from the animal team has been a query about providing plants to allow natural sap feeding by marmosets in the South American House. In their native habitat, sap constitutes an important part of their diet, so could there be some plants to allow this? Here, the easy contact with other zoo horticulturalists proved immensely valuable again. One email was sufficient. Soon a shortlist of possible plants for the outside habitats was drawn up. Actually, it was a very short list – only three plants available in large sizes. The common name of one made it an obvious choice, American sweetgum *Liquidambar styraciflua*. The second was a common hedging plant Portuguese laurel *Prunus lusitanica*. This made sense as there is often very thick resin visible on pruning cuts or insect damage on this. The third was the more common cherry laurel *Prunus laurocerasus*. The Portuguese laurel was the best. The tamarins carved the most intricate patterns in the bark using only their teeth, leaving it looking like a demented junior wood carvers' effort while they collected sap as they went. The poor *Liquidambar* suffered the attentions of macaws that shared the habitat and never did well. A definite challenge, finding a plant that one species of animal would leave alone, yet another could eat.

Prunus lusitanica gently carved by tamarins to feed on the sap.

The native bramble *Rubus fruticosus* is a rampant spreader. New shoots may reach five metres or more in one summer – the tip roots into whatever soil it finds. New growth spreads again from that point next year while the older growth will flower and fruit. It is among the top three plants for nectar and pollen for bees and hoverflies. The berries are a great food source for birds and mammals. It is a very useful plant, great for birds to nest and hide in, but a real thug if allowed to roam freely without control. Low-growing species can be used as very effective ground cover, suppressing weeds. *Rubus* 'Betty Ashburner,' a garden hybrid between *R. pentalobus* and *R. tricolor*, was used on the rhino gabions. It was browsed away entirely at times if growing over the edge of the gabions and down the side, then forgotten. Excellent enrichment – when in reach. One of the parents, *R. tricolor*, is used around the zoo as ground cover. Each week, a few stems are cut to feed the various stick insects in the education centre – which must make it the cheapest food supply for any of the zoo animals.

Willow *Salix* is such a useful plant. There have already been many references to the purple willow *Salix purpurea* used in the gorilla and orangutan habitats as it is not eaten due to the bitterness of the sap from salicylic acid. There are so many stories for the educational team around willow. The medicinal story is the most important, with Hippocrates mentioning it in the 5th Century BC. Willows are great at re-growing, cut them down in winter to near ground level (coppicing), and next year vigorous species will regrow 3m or more in one summer. The straight new stems, which are very flexible when thin, have been used to make baskets or walls as part of wattle and daub construction. There has even been a fishing basket found and dated to about 8,000BC. Many years of use indeed. Modern uses now include biomass production – the quick growth provides a good energy crop from coppiced plantations. The roots can act as biofilters of polluted water. With the quick growth and easy propagation, it is also very useful for land reclamation, for constructed wetlands, soil stabilisation, especially on riverbanks, and as an effective windbreak. Better yet, willows are excellent for wildlife. They attract many species of aphid, which attracts birds and insect predators too. The purple willow and its hybrids are often chosen for much of this work, as they are not eaten by rabbits – which is where the idea of using them in the gorilla habitat came from. Many years ago, poor people even collected the catkins as food – just as the gorillas do at the zoo today. In the chimpanzee habitat, there are several rough clumps of the crack willow *Salix fragilis*, grown from cuttings from the few original trees on the island. This time it was not the gardeners who planted them though. The chimps broke pieces off to eat the bark (which is why the mature trees died) and dragged some pieces around. Being willow, some rooted in where they were left and now form strong clumps that the chimps still harvest shoots from occasionally. They greatly improve the habitat. Introduced into Australia for the same function of erosion control on riverbanks, willows are now an invasive species threatening their ecosystem.

The gorilla habitat gave so many possibilities for planting for enrichment. Most plants were not planted with that intent, though. It was purely serendipity that it happened. Seeing the gorillas gently demolish the overwintering crowns of *Gunnera manicata*, pulling off the protective bracts first to get at the large flower buds in early spring was great, but with the hope they did not damage too much... A little nibbling had been seen in other zoos, but this particular aspect had not been specified. The gorillas were seen pulling up the thick tough roots of bulrush *Typha latifolia* to eat them with a sound like chewing celery. These are edible for humans too, but tough to get out and to clean. In spring, selecting the young flowering shoots of pendulous sedge *Carex pendula* has become a regular pastime, pulling them out and eating just the basal end, which is thicker and no doubt more nutritious. Watching them very gently teasing flowers off prickly *Berberis* shrubs and trying to avoid the thorns, but not always managing to, is somewhat satisfying! Armfuls of clover *Trifolium repens* were collected as they were called in for the night. Picking individual berries from various shrubs – *Lonicera ligustrina* var. *yunnanensis* (*nitida*), *Ligustrum vulgare*, and *Pyracantha cvs*, all give a very limited and slow harvest. Perfect enrichment.

The most intriguing enrichment was with the Asian elephants. A large raised planted area that screened views of the elephant house had been planted with trees, at least 3m out of reach above the elephants. Pieces of timber, leftover stumps about 1m long and 75mm thick, from the birch given as browse, were noticed among the tree branches. How did they get there? The elephants had thrown them there, but why? They learned very quickly that if they did throw a piece of wood up, it was heavy enough to break off a few leaves or twigs which fell down to be eaten. Slow, steady enrichment. No real damage to the trees, though they look a tad rough with dozens of bits of deadwood stuck in them.

Seed Dispersal

There are so many plants around the zoo that form seeds. Examples could be found for many different dispersal systems. As many ornamental shrubs have decorative berries, that is an easy one to discuss. Birds or mammals, such as badgers *Meles meles*, eat the berries and deposit the seeds later – with a little fertiliser.

More unusual seed distribution can be seen with the following.

Cranesbills *Geranium spp* have a distinctive seed head shape, like a crane's beak, elongated and pointed. All the species in the zoo can be used to see this – *G. caffrum*, *palmatum*, and *sylvaticum*. The seeds are held at the base, while the tip of the beak acts as an anchor. When ripe, the end nearest the seed is released, and the seed is literally thrown several feet as if from a trebuchet. Any dead seed head will show the mechanism easily.

Plants have evolved and adapted to many situations, but the most unusual has to be fire – a perfectly natural process in some ecosystems. If the climate is too dry for fallen leaves or twigs to rot, then a fire may be a simple way of cleaning the ground, though it results in nutrient-poor soil. Around the world, there are many areas with plants that can withstand this, with thick bark on trees or tough root systems to grow again if the tree or bush is burnt down. Seeds may be released only after the trigger of the heat of a fire or the chemicals given off from smoke (serotiny). Within the savanna, New Zealand tea tree *Leptospermum scoparium* – a small shrub – has small, very hard seed capsules that exhibit this. They will be held tight until fire triggers them to open, and germination may be better if given an artificial smoke treatment. This is a trick horticulturalists use for such plants – a smoke-impregnated cloth is used to infuse irrigation water with the chemicals in smoke. Most interesting though is that New Zealand vegetation is not usually adaptable to regular fire. Tea trees came from a much drier Australia, which has drier ecosystems, where many plants have this adaptation. Tea tree probably only became widespread in New Zealand as the first Polynesian people started clearing and burning vegetation.

Seeds are often dispersed by wind. A common sight in Ireland is sycamore or ash seeds spinning around in the wind as they fall and travel a little further from the parent tree. The Chinese necklace poplar *Populus lasiocarpa*, planted around the gorilla habitat, has long catkins up to 25cm in early spring, followed by masses of fluffy white seeds, blown around in such quantities it is similar to a gentle snow shower but in summer.

With all berries or unusual fruit, it is always worth mentioning that some can be poisonous, such as yew *Taxus baccata*. Just because a berry looks nice with a glossy red ripe colour does not mean it is safe to eat – check first. With yew, the seed itself is poisonous if chewed, while the fleshy aril is not, which encourages birds to eat them, letting the seeds pass through harmlessly.

Pollination

There can be no subject more fascinating to anyone interested in plants. Pollination is the transfer of pollen from the male part of the flower to the female part. This can be self-pollination (within the same flower) or cross-pollination (with the transfer of pollen between flowers of the same species but on different plants). There are numerous mechanisms to ensure self or cross-pollination, depending on the plant. Cross-pollination depends on the pollen being transported between flowers by wind, water, or animals, but usually by insects. This has resulted in the most wonderful modifications to flower structure. Flower colour (not always visible to us but in ultra-violet for insect vision), the time of flowering (day or night), or methods of attracting insects – not always nectar but includes scents such as rotting meat to attract flies or even mimicking a female bee to attract a male bee. Around the zoo, there are many flowering plants with a variety of pollination systems.

Evergreen Dutchman's pipe *Aristolochia sempervirens*, native to Crete, has a small 30mm flower that hangs down in a U-shape. The opening, with a strong fly attracting scent, encourages small flies to enter. The hole becomes a narrow tube, lined with hairs that are elongated cells and they all point inwards – the fly can crawl in, but not out. On the first day, only the female parts are mature, and the flower can be pollinated if the fly has pollen from another flower on its body. The next day, the male parts mature, pollen is shed, and coats the fly. At the same time, the hairs lining the tube shrivel and shrink – only one cell each – and the fly can escape to the next flower, perhaps to pollinate again. A carefully dissected flower will often have a few small flies inside. All *Aristolochia* are poisonous. This is taken advantage of by some swallowtail butterflies. The larvae eat the leaves, and the aristolochic acid makes them unpalatable to predators.

Aristolochia sempervirens, with tiny flowers that repay a closer look.

Spring sees lots of early flowers, many of which are bright yellow – a colour readily seen by bees. In addition, bees can see ultra-violet, which changes their image of the colour of the flower and often guides them to the nectar they seek. Giant marsh marigold *Caltha polypetala* is a classic yellow spring flower, much planted around some of the zoo lake edges. It is related to buttercups, with a similar flower shape and similar poisonous foliage, so not eaten, but the pollen and nectar are fine for the bees. This and the native marsh marigold *Caltha palustris* are used in various areas along stream sides and lake edges around the zoo.

The African flag *Chasmanthe bicolor* from South Africa has curved flowers in spring – due to their native habitat being a winter rainfall area so growth is in winter. The flower shape has evolved with sunbirds, the old world's equivalent of hummingbirds. As the bird collects nectar, pollen is deposited on the head, ready for transport to the next flower and pollination. Despite there being no sunbirds in Ireland, there is often a good set of seed. Therefore the plants must not be totally reliant on them.

Chasmanthe bicolor with the distinctive curved flower for sunbird pollination.

Another African plant also in the savanna, the red hot poker *Kniphofia*, has similar curved flowers and is often, but not exclusively, pollinated by sunbirds. It is obviously better for the plant not to be tied to any particular pollinator, especially as many plant species generally need to be cross-pollinated to set seed.

The common garden flower *Gladiolus* originated in Africa and now has over 250 species. The pollination strategy varies between the different species. This genus seems to have the most diverse pollination systems. Some are well scented and attract bees and flies. Some have no scent and attract only flies. Others are scented, but only at night and are pollinated by moths. Some use butterflies, and others use sunbirds or beetles. Garden forms are hybrids resulting from years of breeding work, with enthusiasts crossing many different species to get the larger bloom. A variety of colours are available.

Gladiolus tristis flowering well in April, very early and a welcome splash of colour.

Plants move. And not just as they grow. There are many examples of plant movement in response to a stimulus. A tropism is a reaction that depends on the direction of the stimulus. To explain: the growing point of a plant will grow away from gravity (apogeotropism) and towards light (phototropism), whereas roots grow away from light (aphototropism) but towards gravity (geotropism) and water (hydrotropism). Plants respond to touch (thigmotropism). For instance, the tendrils of climbers will often turn towards any surface they brush up against in their efforts to curl over a branch for support. There are exceptions. Some tropical climber seedlings, such as the common indoor Swiss cheese plant *Monstera deliciosa*, germinating on the dark forest floor of dense jungle habitat, head for the darkest spot they can sense, away from any light. This is often the base of a tree between two large buttress roots, and once the seedling hits a vertical surface, it climbs up immediately, heading for light now. Other movements are non-directional, simply a movement in a standard way and these are termed nastic movements. When they respond to touch, they are termed thigmonastic. One example in the zoo would be the stamens in the flowers of any member of the *Berberis* family. Here, best illustrated by *Mahonia x media* 'Charity.' This is a hybrid evergreen shrub that flowers in winter – great for any pollinating insects that venture out if warm enough, such as queen bumble bees. The stamens holding the pollen are held near the petals. When they are touched by an insect or a finger gently, they will move inwards extremely quickly. Look closely, and it is visible to the naked eye. This might be to increase the chance of pollen touching an insects' leg to be taken away for cross-pollinating the next flower.

Mahonia x media 'Charity' is a reliable winter flowering shrub. Look closely, though mind the prickly leaves and you will see the stamens in the centre of the flower.

Nectar is produced by many of the flowers in the zoo but is seldom as noticeable as in *Melianthus major*, also called the honey flower. *Meli* is Greek for honey, as in the honeybee *Apis mellifera*. The red or green flowers of *Melianthus* are held in large spikes at the top of the shoots, easy access for pollinators –

again, often sunbirds in its native South Africa. Copious quantities of nectar are produced, even dripping from the flowers. There is a more sinister side to the plant, as another common name for it in South Africa attests, 'touch me not.' This is an apt name as all parts are extremely poisonous if eaten, including the roots. The plants have a disagreeable smell, which may put off browsing animals.

Persian ironwood *Parrotia persica* grows near the orangutan house. It is wind-pollinated. The ornamental small flowers appear before the leaves in early spring. Many native trees are wind-pollinated, grasses and conifers too. Sometimes bees will collect pollen from wind-pollinated plants, but only if no other better pollen is available. Wind-blown pollen is smaller as it is lighter and has less nutritional value to bees. Wind-blown pollen is produced in vast quantities and causes many people to suffer an allergic reaction.

All brambles *Rubus* are good for pollinators; the native *R. fruticosus* is the very best native plant for nectar and pollen production. Rough ground can often be covered with a twisted mass of vigorous plants, forming a wonderful habitat that is often left alone as the thorns are most effective. With flowers for pollinators, masses of berries for birds and mammals, and a safe nesting area for birds – it is an excellent wildlife-friendly plant, but best kept in a limited area such as a thick hedge. *Rubus* can be found around the world with up to 700 different species, but that becomes very complicated. For instance, the native *Rubus fruticosus* is technically an aggregate. This is a collection of many very similar plants that generally look alike but have slight differences (in the leaves, the prickles, the fruit, or flowers), but not different enough to be a separate species. This arises because the plant is apomictic, which means the seeds may have developed without fertilisation. Resultant seedlings will be identical to the parent plant – a maternal clone.

Back to thorny issues – in botany. Keeping this very simple, the origin of the growth dictates its name: a thorn is a modified stem, a spine is a modified leaf part, while a prickle is a modified hair. So really, brambles have prickles. If anyone wants to be pedantic… The blackberry fruit is not a berry but an aggregate fruit composed of many drupelets. Whereas a banana is a berry…

On the Kaziranga Trail, there are openings in the bamboo forest that are perfect for clumps of *Roscoea purpurea*, a member of the ginger family. In its native habitat, there is only one fly with a tongue long enough to pollinate the flower. This is the suitably called long-tongued fly *Philoliche longirostris*. The fly is found mostly during the main flowering season of the *Roscoea*. When the fly attempts to feed by pushing into the centre of the flower, the pollen sticks onto its back, ready to be transferred to the next flower.

The dark purple orchid-like flowers of *Roscoea purpurea* are very eye-catching.

Physical Defence, Thorns or Gum or Deception

There are many types of physical defence for a plant – stinging hairs, sharp thorns facing forwards or backwards on different parts to make it harder to get in or out, sticky leaves from resin on buds, sticky sap if damaged and on bamboo silica needles on young culms or stems. There is plenty of variety.

A common garden plant around the zoo is *Berberis* in several species and cultivars. The thorns can be effective, but when young, they are still soft. Meaning young growth can be targeted for a while, reducing growth. It was used in the orangutan habitat and left alone by them. Some species were gently eaten one leaf at a time by the Siamang gibbons. This took some time, sitting next to a small plant, carefully picking just one leaf each time, constantly going back for another. The flowers are keenly visited by bees and hoverflies, an excellent pollen and nectar source. As in *Mahonia* above, the stamens are thigmonastic and move when touched if you look very carefully.

Not all defence is a simple thorn. The anchor plant *Colletia paradoxa* (discussed in the savanna above) has the most unusually shaped stems, hence the common name. The sharp points are technically cladodes – a flattened shoot, here they are very stiff and rigid, making navigating through the plant a challenge. The flowers are minute, but pleasantly scented and are well pollinated as each flower develops into a three or four seeded pod that is easy to collect – if you have pliers to reach in and grip them. This unusual-looking plant is in the *Rhamnus* family and has symbiotic bacteria living in its roots, increasing the nitrogen available for growth.

A more normal thorny defence is found on *Gleditsia caspica* in the savanna habitat. It has short spines with even tinier spines on those spines, nothing very big to see, but spikes in every direction. The plant looks a little like an *Acacia*, with typical legume leaves – hence the planting in the savanna habitat. Within the Education Centre, there are dried branches of *Acacia karroo* from plants grown in the Palm House at the National Botanic Gardens, Glasnevin. The thorns are enormous, maybe 50mm long.

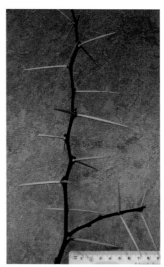

Acacia karroo has remarkably long thorns

The most odd-looking plant on the savanna and very deceiving is *Pseudopanax ferox*. It has a few upright branches and very strange, narrow, spiky, hard, stiff brown leaves, up to 300mm long. It has one of the most intriguing stories for the education team. After ten years or more, once the plant becomes mature, the leaves change. They become shorter, much wider, and a normal dark green colour. It is believed this is an evolutionary adaptation to avoid being eaten. Now, remember what animals are native to New Zealand... It was dominated by birds for millions of years. In fact, the first European visitors were staggered by the dawn chorus. One of the largest birds was a moa. This stood 3m tall and browsed vegetation. So, imagine a young *Pseudopanax ferox* trying to avoid being eaten. Make the leaves brown and they look dead, unappetising. Make them stiff and rigid, harder for the moa to break them off or bite through them. Then, once they gain enough height, make the leaves bigger and better. Very sneaky. The moa, hunted by the Maori, are unfortunately extinct, but the *Pseudopanax* has not realised that yet. There are several other *Pseudopanax* species around the zoo, with a very distinctive flower head that looks like ivy *Hedera*. They are the same family, *Araliaceae*. The scientific name means simply false (*pseudo*) ginseng (*panax*). *Panax* is a related genus in the same family. The specific epithet *ferox* is simply fierce or thorny, referring to the hard leaves.

Pseudopanax ferox with stiff downward facing leaves, unique appearance.

A very common garden shrub is *Pyracantha*. Available in several cultivars with different coloured berries. It is unpleasant to work with sometimes and a great deterrent for a garden wall or hedge. Birds frequently nest in *Pyracantha*, good rigid branches and a thorny barrier too. Perfect. White flowers in spring are excellent for pollinators – like most members of Rosaceae, the rose family. In autumn, the berries are eagerly sought after by birds, but only if red. Birds select the red berries first as they look ripe and maybe are easier to see against green foliage. Orange berries would be second choice and yellow last – are they ripe or not? A useful plant. So many wildlife links.

The Ichang bramble *Rubus ichangensis* is a delight. The prickly defence is particularly fiddly, as the stem has prickles facing one way, while the leaf stem and the leaves have them facing back – catching an unwary hand both ways. Another excellent wildlife plant. The small flowers are incredibly good for pollinators.

Rubus ichangensis covered with small flowers and lots of pollinators.

The ultimate thorny defence has to be gorse. A double form *Ulex europaeus* 'Flore Pleno' is the only form planted in the zoo. The reproductive parts have been replaced with petals, hence the double form. With extra petals there is no seed – a good thing too as ordinary gorse can be a very vigorous weed, seeding in exactly where not wanted. Seeds shoot out up to 3m when ripe. The pod twists and squeezes the seed out under pressure. The whole plant is viciously thorny on all surfaces. Excellent wildlife plant. Though there is no nectar or pollen with the double form, making it less useful. It makes a great safe nest area for birds. Within the zoo, it is used as enrichment for the gorillas and chimps. Both carefully take the flowers to eat, one at a time. Great enrichment.

Botany, the Study of Plants

The study of all aspects of plants is botany. What a wonderfully varied subject. Presenting botany to the education team was a challenge, purely because there is so much information, so many species, so many stories, some more complicated than others. Where to start is as difficult as where to stop. The examples below are a taste of what can be used for educational tours or classes. There are many more around. This is a very eclectic mix of facts about many plants around the zoo, no real order...

Plants – and animals – often have a very defined design that can be represented by mathematics. Very often this follows the Fibonacci Sequence of numbers 0, 1, 1, 2, 3, 5, 8, 13, 21, 34, 55. Fibonacci was an Italian mathematician in the early 1200s and wanted to calculate how many rabbits he would have after a year if starting with two. Each number is the sum of the previous two numbers. The sequence can be applied to a lot of shapes and structures within nature. Look at a sunflower seed head, the seeds are in spirals. The number of spirals and the number of seeds in a spiral, are both numbers in the Fibonacci Sequence. Other examples include pinecones such as the Korean fir *Abies koreana*. The number of spirals and scales per spiral fit in the sequence. A snail shell will fit. By drawing the shell as diminishing squares, the length of the sides will be in the sequence.

Chinese gooseberry *Actinidia deliciosa* may not be familiar to everyone, but kiwi fruit certainly is. Here we have marketing within botany. Many years ago, farmers in New Zealand started growing *Actinidia*. They taste and look a little like gooseberries, and they came from China, hence Chinese gooseberries. Commercially, the marketing department thought the name could be a hindrance and as they were being grown in New Zealand thought up the name kiwi fruit, named for their national bird the kiwi *Apteryx*, which is furry looking and brown too. A rampant climber to 10m, (the plant, not the bird) with large leaves, with male and female flowers on different plants, so both need to be planted or no pollination, no fruit set. Within the zoo, it is a regular diet item for many primates. It can self-seed too, as was discovered when one seedling grew halfway up a wall in an old chimpanzee exhibit, lodged in a crack with a dollop of fertiliser thrown by the chimps, but no soil – it never grew more than 300mm long. Another self-sown seedling at the South American House grew very well, twining through the steel mesh, nicely camouflaging the habitat and giving a great natural climbing structure. Very bouncy, for the tamarins, encouraging natural behaviour. The five freedoms again.

In the savanna, there are several *Acacia* species which have a fascinating educational handle for the education team. For simplicity, *Acacia* has been used throughout this book, but recent name changes mean that many African species are now classified as *Vachellia*. In nature, there is a constant battle between animals wanting to eat plants and plants resisting. *Acacia* has three defence mechanisms. Within fifteen minutes of being browsed or nibbled by any animal, many *Acacia* produce higher levels of tannin in their leaves, which combines with proteins in the gut of a giraffe (or other browsing animals) to make the leaves indigestible. At the same time, they produce ethylene which wafts downwind and alerts neighbouring *Acacia* trees to the danger, and they then produce more tannin in their leaves. Soon after starting to eat *Acacia*, giraffes have to move on – upwind. This obviously limits damage to any one group of trees. There has been high mortality of kudu *Tragelaphus* (a large antelope) in some large commercial game ranches as food ran short in the dry season. Despite the tannin, they kept eating the *Acacia* as the only available food.

Tannin is a common substance used by plants to deter animals or insects from eating leaves or bark. Think of oak *Quercus* leaves, a dark red when young from tannin content to reduce caterpillar damage. Though tannin is useful to us, harvested from oak bark as a curing agent in leather production (tanning). The most obvious tannin is found in tea, giving the brown colour to most teas and a little bite of bitterness. The picked tea leaves are the youngest two leaves or so, and, as in the oak tree, to protect young growth more tannins are produced.

But *Acacia* do not stop at that. They have incredibly long thorns on many species as a physical defence. Giraffes get around this by having very long tongues, which can get between the thorns and strip the leaves. The plant is also sneaky. Thorns cost energy to grow. They are usually only produced where needed, around the outside edges of the canopy. The centre of the tree canopy, where usually no animal can graze, often has far less.

There is one even sneakier trick by the plants. Some *Acacia* species have hollow thorns, which are occupied by Acacia ants *Pseudomyrmex ferruginea*. This ant is only found living in *Acacia* tree thorns. If a giraffe (or other browsing animal) attacks that plant, the ants come out to bite noses or whatever and scare the animals off. Nasty. But it is much more intriguing than that. The trees produce a small lump of food for the ants – proteins and sugars – called Beltian bodies (named for Thomas Belt, an English geologist who described this symbiotic relationship) on the tip of each little leaflet. The ants feed their larvae almost exclusively on these little protein parcels, and the tree gets protection from most other insects and herbivores. Neat.

There is a downside to *Acacia*. It is remarkably tough, grows well in poor soil with little water. Being a legume no doubt helps, with beneficial bacteria in the roots providing nitrogen. The species growing in the zoo often grow too quickly, are too well fed and watered, and branches break off regularly. The reliable toughness and fast growth certainly encourage their use for timber or paper production in various parts of the world. Where they are dry and warm enough, they grow well. Unfortunately, they flower and seed well too. They self-seed a lot,

Fascicularia bicolor happy in the fork of an oak tree.

The small blue flowers are only a few mm across, but the red leaves catch the eye.

One of the most intriguing families has to be *Passiflora*, passion flowers. The flowers are most complicated and decorative, in various colours, shapes, and sizes depending on the species – there are over 500. The name comes from Latin *passio*, meaning passion, and *flos*, meaning flower. Found around the world in warm climates, they are much more diverse in South America. Usually seen as climbers, some are actually trees. The name originated from Christian missionaries when they first landed on the continent, picked on the leaf and flower shape and tendrils as symbolic of the crucifixion. The leaves of many species have five leaflets, like a hand (the hand of the oppressor to the priests) while the curling tendrils represent whips. The ten petals and sepals of the flower reminded them of the ten disciples without Peter or Judas. The three stigmas recall the three nails and the five stamens, the wounds.

Botanically, the genus has so many interesting points. They are generally climbers, with tendrils attaching to anything. Tendrils start straight, twisting around in a search for support – called circumnutation. If they stroke against a twig, they twist rapidly around that. Once anchored at the tip, the tendril then bends into several coils in opposite directions, with a short straight break between – this is called tendril perversion. Once this has formed, the coils rotate against each other, tightening up the whole support system, and then it becomes woody to stay firm for longer. Charles Darwin was fascinated by tendril movements.

Inspect the leaves. There will often be small lumps under the leaf or maybe on the leaf petiole, as in *Passiflora caerulea*. This is the commonest species grow. in Ireland, with a few around the zoo. These lumps are a remarkable adaptation to butterflies, especially *Heliconius* species – the lumps mimic the butterfly eggs and deter some egg-laying. Each *Heliconius* species specialises in a particular *Passiflora* species, so the lump size, shape, and texture varies. Some *Passiflora* species have sticky hairs on various parts of the plant that trap and kill small insects, from which the nutrients are then absorbed. Not a true carnivorous plant, but certainly a useful extra food supply. *Passiflora* pollination is often linked to a specific insect – or bird. Many species are only pollinated by the sword-billed hummingbird *Ensifera*, whose bill is longer than the body.

The flower of *Passiflora caerulea* is stunning enough, but the stories that can be woven around the plant are better yet.

Around the zoo, *Pseudopanax*, *Fatsia*, *Aralia*, and *Hedera* all have the same flower structure, just size difference. It is very easy to see the family Araliaceae relationship. Such different plants, though. Common ivy *Hedera helix* is a rampant climber that clings to any surface and only flowers once high enough. The leaves and the shoots change shape as well, as no more adventitious roots are needed. Cuttings taken from flowering wood will not creep but remain as woody shrubs. Ivy is a very good wildlife plant. Early autumn flowers attract pollinators, while the berries feed many birds. Dense growth is great cover for bird nests and insects hibernating. Before it climbs, it may be a simple groundcover plant, and there it is not so good, easily swamping all other growth with its waxy, evergreen leaves and forming a monoculture. The seeds, spread by birds, germinate far too readily, often at the base of shrubs or trees. Ivy is one of the worst weeds, not easy to control.

Pseudopanax has many very choice evergreen shrub species, with wonderful large palmate leaves. Several have a similar leaf variation to ivy, with juvenile leaves very different from the adult. The most often seen of these is *Pseudopanax ferox*, whose young leaves are brown (they look dead so are not eaten), very hard (so less palatable), and spiky (again less palatable). For the education team, remember it is native to New Zealand, with no native mammals that would eat it. It is believed the now extinct flightless moas were the main herbivore, and their heavy sharp bills would have been able to cut off small branches and leaves easily. In *Pseudopanax ferox*, the leaves become larger and fleshier when the tree is about ten to fifteen years old, and the top is above the height of the tallest moa, so safe from browsing.

The sessile oak *Quercus petraea* is one of the best native Irish trees for biodiversity – insect life, birdlife, lichens, fungi, algae – especially on old trees, with hundreds of species found living in this vertical habitat. New leaves in spring are red, with high concentrations of tannin, which makes the leaves look dead (not nice fresh green) and less attractive and tastes bitter, so less palatable. This reduces the damage, but the tree compensates for leaf loss by putting out secondary growth in summer – lammas growth, named for August 1st, a summer festival, and harvest date. The main spring flush of growth triggers so much insect activity that insectivorous birds time their chick-rearing to coincide, but changing weather, early or late springs, can be a problem.

Common lime *Tilia x europaea*, a natural hybrid between the small-leaved lime *T. cordata* and the large-leafed lime *T. platyphyllos*, is a very useful tree. It is quick-growing easily from cuttings and is a favourite of aphids. That is great for insects or birds that eat aphids, but the aphids produce incredible amounts of honeydew. This is waste from the aphids, as they have to take in so much sap to get their nutrition that surplus liquid is excreted in quantity. The sap pressure in the plant will actually force the sap through the aphid too. Honeydew contains enough sugar to attract ants, and even honeybees collect it. It is very sticky as it dries – not a great tree to park a car under. The sugar content may encourage fungal growth – sooty mould – a black layer that looks awful. The flowers are very good for pollinators, a great nectar source, and very good for honey. Each tree is said to be worth an acre of clover. The tree itself is a good source of stakes, firewood, and wood for turning as it does not warp. If coppiced, the bark used to be made into rope. The name has nothing to do with the lime fruit, which is from the citrus family. *Tilia*, literally from the Latin name for this tree, is known as linden in many parts of Europe, including Sweden, where *linn* is the root of the surname of Carl Linnaeus, who everyone knows as the father of modern taxonomy, using binomial nomenclature.

There are two Brazil nuts in the Discovery and Learning Centre – not the individual seeds that are sold, but entire seed cases. Though called Brazil nuts, they are found throughout much of the Amazon rainforest. Each heavy woody seed case is about 125mm across, circular, and very woody and each has been opened differently. Only two animals on the planet can open them; humans, with a heavy knife or axe, and agouti *Agouti dasyprocta* (a large rodent) using its teeth. For the tree to grow, the seed has to be taken out of the woody shell. The agouti does this, eating some seeds but burying others for later, a food store in times of need. Not all the buried seeds are found, though, and they are spread over a wide area, thus ensuring new young trees. Not too many are needed, as Brazil nut trees, *Bertholletia excelsa*, can live easily to 500 years, reaching 50m tall, one of the tallest emergent trees in the rainforest. Commercial collection of the nuts can be too good, with not enough left to grow.

Brazil nut opened with a heavy bladed machete.

Brazil nut opened by an agouti – teeth marks very visible.

With such a valuable crop, plantations have been tried but with little success. This is due to poor pollination, and there is a great storyline on this. The only insects that can pollinate the flower are large-bodied *Euglossid* bees. The females pollinate the Brazil nut flower, a large heavily petalled flower that is difficult to get into. The female bee needs a mate. She will only be attracted to a male *Euglossid* bee if he has the right perfume he must collect while pollinating an orchid, *Coryanthes*, which does not grow on the Brazil nut tree. There needs to be a mix of trees, and the best would be a pristine rainforest.

So, the Brazil nut tree needs *Euglossid* bees for pollination. The bees need other trees to grow the orchid, or they will not breed. The tree then needs agoutis to open the seed case and hide the seeds. The empty seed cases, tough and woody, often collect rainwater, and in the Para region of Brazil, tadpoles are often found of the Brazil nut poison frog *Adelphobates castaneoticus*, which uses small pools of water such as in epiphytes described above. The specific name *castaneoticus* comes from *Castanea*, the sweet chestnut of Europe, as Brazil nuts were called chestnuts when first found by Europeans.

While looking at diversity, mention must be made of fungi. They are everywhere, every day, essential, though not always welcome. Current estimates range from 100,000 species to 1.5 million – the disparity is the usual reason, unfortunately. There are fewer taxonomists in the tropics, where more species of almost anything are found. Mention fungi, and many people will only think of the mushrooms they can buy in shops. Think of other fungal food – yeast for bread making is a fungus, as are the yeasts used for wine or beer. Many antibiotics are fungal based. Think penicillin. Not all fungi are nice for us though, consider athlete's foot…

Fungi are natural partners with the vast majority of plants, a symbiotic relationship again, with the fungus extending a vast mycorrhizal system, looking like cotton wool in the soil. The plant gains from better, more extensive food and water gathering, and the fungi obtain carbohydrates from the plant. Some fungi kill trees and shrubs, but there are beneficial fungi that help the plants resist. There are now commercial suppliers of beneficial fungi that greatly assist plant growth and establishment, especially in poor soils.

Not all fungi are beneficial. Some are active killers of large trees, slowly infecting the timber, quietly and unseen. Until a storm breaks off large branches or topples a tree, and surprise, only a third of the wood is still good, the rest having turned to sponge with no strength. It is all part of the natural cycle of growth and decay. However, it is always sad to lose a good tree.

Bracket fungus *Ganoderma* on beech *Fagus sylvatica*, at 8m above ground, with no easy accurate way of checking degree of decay. Worrying. Hence the decision to fell. Almost the entire tree was felled in two sections and laid down. It was used as natural climbing material on a nearby primate island.

Some fungi have an incredible symbiosis with algae or cyanobacteria to form lichen, which are not plants, but have some plant-like qualities. Lichen is incredibly tough, surviving in many extremes, and is often the first coloniser of bare rock. Incredibly slow growing and hard to correctly age, but estimates of 8,000 years make them potentially the oldest living thing on the planet.

Several different species of lichen on tree bark.

Fungi make up the most significant amount of biomass within the soil. It gently breaks down the cellulose of dead plant or animal material, a vital part of the natural cycle. The zoo mulch would often be held together by masses of white fungal mycelia, gently rotting down the tree chips. The mulch attracts masses of the common fluted bird nest fungus *Cyathea striatus*. This is literally like a very small nest with four or five 'eggs' that are the spore bodies of the fungus – spread by splashes of rain. Like most small fungi, seldom seen, but present everywhere when looked for. Many other fungi pop up through tree and shrub-planted areas. Most are harmless, some very unusual, and all are interesting in their own way.

Fluted bird nest fungus *Cyathea striatus* forms large mats of 'nests' – each 'nest' only to 10mm across – on the wood chip mulch. The small 'eggs' carrying the fungal spores are thrown out of the 'nest' in heavy rain, spreading the fungus.

Crested coral fungus *Clavulina coralloides*, to only 75mm, is very frequent and just loves the deep mulch.

Gymnosperm or Angiposperm

One very basic division in the plant world is the way seeds are produced. The division gymnosperm literally means naked seed, as the seed is not enclosed in an ovary but is exposed and often held in cones. This group includes conifers (the largest number), cycads, gnetophytes, and *Ginkgo* – just one species now. In contrast, Angiosperms have an enclosed seed, but more importantly, they have a more noticeable flower. Hence they are called flowering plants.

Many flowering plants have already been included in various areas around the zoo, but the following gymnosperms have also been used and can be interesting educational points.

The maidenhair tree *Ginkgo biloba* is a living fossil, with no other species in this family and unique fan-shaped leaves. Fossils have been found that date to 270 million years ago. A fossil was sourced for the Discovery and Learning Centre showing a leaf of a long-extinct *Ginkgo* species, looking similar to the modern living leaf. *Ginkgo* is a long-lived tree to maybe 2,000 years, many chemicals in the wood resist fungi and insects. At the atom bomb site at Hiroshima, six *Ginkgo* trees survived the radiation blast despite being within 2km of the centre. Several *Ginkgo* have been planted around the zoo.

Another extraordinarily ancient plant grows in the savanna, *Ephedera gerardiana*. This is native to a wide area around the Himalayas, other species have been found around the world. Growing to only about a metre tall, it is a mass of green twiggy growths, not eye-catching but very distinctive. Fossils have been found dating to around the early Cretaceous, about 150 million years ago – they would have seen the dinosaurs develop and disappear. The flower and seed are more like that found on a conifer.

Ephedera gerardiana in the savanna, with a conifer look to it, gently spreading.

Rough horsetail *Equisetum hyemale* is another dinosaur age plant. Native across much of the northern hemisphere, this was brought in labelled E. camtschatacense, suggesting it might have been introduced from Kamchatka in Siberia. Quite possible. It is certainly tough. Snow and frost do little damage, an evergreen herbaceous vertical gem. Each thin green stem has ribs along the entire length and nodes circling at intervals. Sometimes at the top, there is a tuft of brown that produces spores like ferns. Growing to about 1m, it is only used in one part of the zoo – in the middle of the stream through the Kaziranga Trail, between concrete edges that hold water. The simple reason is the roots. Horsetails of several species can be the most obnoxious weeds once they are established. Delving deep and spreading far, each piece of root broken off while digging will grow again – the easiest way to be rid of them is apparently to move house. Extinct species grew to 10m – imagine that as a weed. But think of a forest of those 10m stems, looking like a clump of tropical bamboo, growing in marshy conditions, with dinosaurs strolling through – what a picture. Were they eaten much? The stems are so full of silica that they may have been less palatable. The silica makes the stems so tough that Stradivarius used them as sandpaper to polish his violins.

A *Calamites* fossil, related to *Equisetum*, in the Discovery and Learning Centre. Some *Calamites* could grow to 30m.

Equisetum hyemale in its concrete containment area. Note at the top of some shoots the black parts are fruiting bodies, like miniature cones.

Plants are very obvious in our lives, but there are other groups of organisms that are almost but not quite totally unlike plants as they can also photosynthesise – algae and cyanobacteria. Most are green. Some are brown (seaweeds). Algae is a loose term for many different families, the largest being the giant seaweed kelp *Macrocystis pyrifera* that can grow 500mm per day to 45m tall. The smallest *Prochlorococcus*, a type of cyanobacteria (or blue-green algae), was only discovered in 1986. No wonder it is so small that many thousands would fit in a mere millilitre of seawater. Numbers really count though and estimates are that this tiny organism is responsible for up to 20% of atmospheric oxygen. Looking back at the time span of earth, as mentioned in the introduction, it is estimated that the oxygen production from these tiny cyanobacteria, as a 'waste' product of photosynthesis, slowly accumulated over a billion years to form an atmosphere that was breathable for animals. Such a tiny organism is at the very base of food chains.

Every time that algae in the zoo lake becomes a problem, it is worth thinking about and reminding people of the debt we owe it for the very air we breathe. More developed plants carry their own cyanobacteria with them in the form of chloroplasts, which are the organs that conduct photosynthesis in the plant. Each chloroplast has its own DNA, because they have evolved after photosynthetic cyanobacteria was engulfed by a different form of organism, an early plant. The result was so successful, helping each other, that they developed into the plants we have today.

There is another storyline on cyanobacteria in the zoo – the links to specific animals or higher plants. *Gunnera* has already been mentioned as having a unique association with cyanobacteria between the cells with the usual symbiosis – the *Gunnera* providing fixed carbon and the cyanobacteria receiving nitrogen. Talk about this near the flamingo habitat, and there is another storyline with the flamingos. Why are they pink? Their natural food (found in the water) is a varied mix that includes cyanobacteria. Each flamingo species is a slightly different colour, depending on its natural diet. Those that feed on cyanobacteria have the darkest colouring. It is from the carotenoid canthaxanthin present in the cyanobacteria. The last storyline has to be the strangest, and visitors need to go to the South American House where the two-toed sloth *Choloepus didactylus* are kept. Imagine these famously slow mammals, in their rainforest habitat, high up in the canopy, where they are preyed on by harpy eagles *Harpia harpyja*, the largest of the rainforest raptors. To help disguise the sloth in the trees, the fur has a special groove in each hair, in which cyanobacteria can grow and form a green covering over the fur, blurring the sloth into the rainforest. There can be so much cyanobacteria growing at times that the sloth can actually graze it off to eat.

Conifers

Conifers (trees that bear cones) are such a vital plant group. Many have already been talked about as part of the landscaping for the sea lion cove where conifers fit in as habitat vegetation. Rather than repeat too much, consider conifers per se for the education team. The largest forest on the planet, a wide belt around high latitudes in the northern hemisphere, is the taiga or boreal (northern) conifer forest. There are different genera across the forest, only to be expected over such a wide area. Squeezed between the tundra further north and the temperate forest further south, many conifers adapt perfectly for cold winters and heavy snow. Narrow canopies and drooping branches shed snow with little or no damage. Evergreen conifers fare better in cold climates as the soil is generally more nutrient-poor – the soil is so cold that organic matter decays slowly. Thus, fewer nutrients are available. If too cold, the leaves of evergreen broadleaf trees suffer, whereas conifer needles are far more cold-tolerant.

Almost half of the timber produced around the world comes from conifers. Many conifers are from the southern hemisphere, often in very limited ranges, maybe on islands, and at great risk. The Wollemi pine *Wollemia nobilis* is probably the best known critically endangered conifer. Only discovered in 1994 in hard-to-access sandstone canyons in New South Wales, Australia, this is similar to the *Ginkgo* above – a living fossil. Only about 100 trees survive here in three groups, but many have been propagated and distributed worldwide. The trees are very similar genetically, and this poses another issue. The introduced invasive fungal disease *Phytophora cinnamomi*, a serious problem in many parts of Australia (and other countries too), has been detected in the area. With a limited genetic pool, all could succumb. Equally, widespread wildfires could do the same in a short time.

Conifers are found worldwide in diverse forms and grow in very varied climates, from hot and wet to dry and freezing – and everything in between. They are also record-breakers. The oldest bristlecone pine *Pinus longaeva* is an estimated 4,700 years old. The tallest coast redwood *Sequoia sempervirens* is 115m. The largest giant sequoia *Sequoiadendron giganteum* is an estimated 1500m^3 of timber. Conifers would have made up a large part of the diet of herbivorous dinosaurs. The need to reduce such browsing would explain the wide range of chemicals and resins present in many species – a very early plant defence system. Within the horticulture world, many garden-worthy plants have been selected for many reasons. These include a particular growth habit, may be very narrow fastigiate (upright), good-looking silvery foliage or seen the most, a dwarf habit where the plant stays very small or simply grows very, very slowly.

Conifers as a group are high on the list of endangered species, with at least one-third of the 600 species listed, many as critically endangered. Though often seen in gardens, many are at great risk in their native habitats. Reasons include too much felling for timber extraction, but more often, the threat is from invasive insects or fungi, sometimes causing thousands of trees to die. Non-native invasive pests and diseases are very difficult to control once in a habitat.

Unfortunately, it can produce thousands of seeds which are a nuisance at times. It has been introduced into America and New Zealand, where it has become invasive.

Lyme grass *Leymus arenarius* is planted near the sea lions. It loves growing in sand. There is a name for that – psammophilic. It literally means 'sand liking.' An interesting grass in its own way, it is very tolerant of saline conditions as it has adapted to coastal living. It is not just coastal anymore, it likes some roadside verges where salt is applied each winter to prevent ice on the roads, which spreads to the edges where most plants would be killed or stunted – not *Leymus*. Very happy there, thank you. Compared to other grasses, it has more fungi associated with its roots. This greatly helps water and food collection in a dry and hungry root zone.

Silver grass *Miscanthus sinensis* was used in the savanna, but it has a giant relative, *Miscanthus x giganteus*. This is a hybrid of *M. sinensis* and *M. sacchariflorus*, and most importantly, it is sterile, so no seeds are being spread around. It has typical hybrid vigour, and is worth listing as it is potentially a valuable energy crop, growing up to 4m tall, cut down each winter. This is a highly efficient C4 grass, possibly the highest efficiency of any biomass planting, twice as good as willow *Salix*. The stems can become sawdust-free bedding, an excellent option for horses where dust can be a health issue.

Bamboo has to be the most useful grass, with an estimated 1,000 uses, particularly throughout Asia. Young shoots of some species can be eaten. Mature culms are tough and woody but easily split into long thin lengths to make baskets, mats, hats, or umbrellas. Thick diameter culms become posts, walls, furniture, tools, musical instruments, or paper. The thickest tropical bamboos can have a culm (stem) 25cm thick at the base, cut horizontally to make buckets or pots. Look at village life in many areas, and everywhere is the ubiquitous bamboo in different forms. It would indeed be hard to imagine village life in Asia without it.

RESOURCES

There are endless gardening books detailing how to grow, propagate or prune every plant imaginable. If they have an emphasis on the plants, many travelogue-style books, would be useful for a general 'feel' of natural vegetation in a given area when deciding on a planting list for a new, themed project.

In addition to a extensive selection of those, the following books proved invaluable, detailing the poisonous issues or an educational aspect of a plant that is never thought about normally. Appreciating a plant for its natural abilities, soil formation, the diversity it encourages, water cleaning – or whatever, should be a more regular part of our thinking towards plants.

A Dictionary of Useful and Everyday Plants. F.N Howes. Cambridge University Press 0-521-08520-9. Published 1974. Masses of useful information, especially about the uses of plants and their families.

Stearn's Dictionary of Plant Names for Gardeners. William T. Stearn. Cassell ISBN 0-304-31449-5 Published 1992. I can only repeat a quote on the fly leaf by William Cole in the 17th Century in respect of the gardener's particular pleasure 'to have plants speaking Greek and Latin' to him and putting him in mind of stories which otherwise he would never think of.

Plants for People. Anna Lewington. Eden Project Books. ISBN 1-903-91908-8 Published 2003. An inspiring book, with an incredible breadth of information.

Green Inheritance. The WWF Book of Plants. Anthony Huxley. Gaia Books. ISBN 1-85675-000-0 Published 1984. An incredible resource, full of useful information. Every teacher and politician, should have to read it.

Mind-Altering and Poisonous Plants of the World. Michael Wink and Ben-Erik van Wyk. Timber Press. ISBN-13; 978-0-88192-952-2. Published 2008. The main go-to book for any plant poison question. Brilliant resource. And with that title, always good to leave on the desk...

Horticultural resources within the European zoo community developed dramatically in the past thirty years. Peer support from zoo horticulture colleagues across Europe has been inestimably valuable. The plant database on www.zooplants.net, a joint effort between the Zoological Society of London www.zsl.org, the British and Irish Association of Zoos and Aquaria www.biaza.org.uk, the European Association of Zoos and Aquaria www.eaza.net and Zoolex www.

zoolex.org is a remarkable achievement. Like the plants it details, it is continually growing.

Zoolex itself is a wealth of information, especially the hundreds of detailed descriptions of many zoo animal habitats, with a careful record of fences and buildings and special features to improve animal husbandry, specifically for the animal species held there. Planting details are also included, usually with a complete plant list, a great way of seeing how different plants have survived or not.

Of course, at all times, the plants are just one element to help with the successful conservation of animals. For many animals this is of great benefit. They are not always under observation from all angles, which helps satisfy some aspects of the 'five domains' or 'five freedoms' of animal welfare. These started as goals for keeping animals on farms in the UK and are worth repeating in fuller form. The World Association of Zoos and Aquaria www.waza.org has upgraded the concept, prioritising the need to keep animals in the best possible environment within the zoo setting.

1. Freedom from hunger and thirst – by ready access to fresh water and a satisfying diet to maintain full health and vigour.

2. Freedom from discomfort – by providing an appropriate environment, including shelter and a comfortable resting area.

3. Freedom from pain, injury, and disease – by prevention or rapid diagnosis and treatment.

4. Freedom to express normal behaviour – by providing sufficient space, proper facilities, and the company of the animal's own kind.

5. Freedom from fear and distress – by ensuring conditions and treatment that avoid mental suffering, including screening from other animals or visitors.

INDEX

Please note, the index is listed by scientific name for the plants, but by common English name for the animals or insects.